CONTEMPORARY NEWTONIAN RESEARCH

STUDIES IN THE HISTORY OF MODERN SCIENCE

Editors:

VOLUME 9

CONTEMPORARY NEWTONIAN RESEARCH

Edited by

ZEV BECHLER
Dept. of Philosophy, University of Haifa, Israel

D. REIDEL PUBLISHING COMPANY
DORDRECHT : HOLLAND / BOSTON : U.S.A.
LONDON : ENGLAND

Library of Congress Cataloging in Publication Data

Main entry under title:

Contemporary Newtonian research.

(Studies in the history of modern science ; v. 9)
Includes index.
1. Newton, Isaac, Sir, 1642–1727. 2. Physics—History. 3. Science—History. I. Bechler, Z. II. Series.
QC16.N7C66 509′.2′4 [B] 81–19991
ISBN 90–277–1303–0 AACR2

Published by D. Reidel Publishing Company,
P.O. Box 17, 3300 AA Dordrecht, Holland.

Sold and distributed in the U.S.A. and Canada
by Kluwer Boston Inc.,
190 Old Derby Street, Hingham, MA 02043, U.S.A.

In all other countries, sold and distributed
by Kluwer Academic Publishers Group,
P.O. Box 322, 3300 AH Dordrecht, Holland.

D. Reidel Publishing Company is a member of the Kluwer Group.

Printed in The Netherlands

TABLE OF CONTENTS

Z. BECHLER

INTRODUCTION: SOME ISSUES OF NEWTONIAN HISTORIOGRAPHY

The only way to reconcile the big-time folly of the 17th century with our 20th century small-time coneption of science is to minimise the failure and maximise the achievement of that century of genius. The humanistic effort of post-Kuhnian philosophy never really changed anything in this old basic mental strategy, which began its work even in Newton's life with his first apologetic disciple, Roger Cotes, and went on uninterrupted through the 18th century Enlightenment idolators, through the 19th century positivists, and up to our own Newtonian-scholarship Industry. The modes of approach to the 17th century genius were always well established even before the approach was made, and could be pretty nearly predicted. Thus, even though there would be always a straight-laced Strong [11] to attack a woolly minded Burtt [1] on the issue of the relevance of metaphysics to science and mathematics, the attack would somehow miss the point, for "relevance" would for ever remain too misty a concept to get any real work done. Burtt, or F. Yates, would never even bother to argue that some metaphysics is a *sine qua non* of any working science and scientist, but they would not have to bother – everyone accepts this. And again, M. Hesse, [5] retracting Strong's strategy, would vehemently defend the autonomy of science, yet who would care to question this?

On the other hand, look at the positive side of the arguments. Burtt opened his book with a question about the acceptance of Copernicus' astronomy when nothing palpable was going for it and almost everything was against it, and his answer could only be in terms of metaphysics by the definition of the game, since he assumed from the start that any scientific answer was inadequate. Here was a question about the logic of acceptance of theories by scientists and it assumed that scientists, just because they are rational, must depend on metaphysics to make big decisions. The holistic implication of conventionalistic historiography from Duhem and LeRoy through Cassirer and Koyré, right up to Kuhn and Feyerabend, never abandoned this belief in the deep rationality of science and scientists. There was no other way to be rational, so they presupposed and argued, than to employ a whole conceptual system, a paradigm, a Dogma – the whole person, if it comes to it – in order to accept and reject scientific assertions. And if the

Z. Bechler (ed.), Contemporary Newtonian Research, 1–20.

outcome is that the only way to be rational is to be irrational what of it, as long as it is rational?

This belief in the overall rationality of the scientist is the unbreakable tie that unifies historians of scientific ideas into one big loving family in which disputes can't really be fundamental. Here everyone works for the common good and deviations are negligible, and the common good is an exhibition of true rationality silently throbbing wherever science exists.

A major effort in this direction is Cohen's explication of what he calls the "Newtonian Style," namely, the distinction which Newton presumably drew between the domain of mathematics and the domain of physical reality. Any failure to understand this Newtonian style, Cohen states, is bound to rob us of the possibility of truly "appreciating the power of Newtonian methods and the full scope of his achievements" [3] p. 22.

In particular, Cohen argues, regarding this Newtonian style as the essence of the Newtonian revolution will clarify, among other issues, the dispute about the role of Hermeticism in the formulation of the concept of universal gravitation and the concept of forces acting at a distance. The problem about the concept of forces acting at a distance is solved as follows:

Newton could proceed with his work on the *Principia* without any commitment to the existence and nature of forces, because "From the point of view of the Newtonian style, such investigations (phases one and two) could be undertaken without the question ever arising of Newton's commitment to the independent existence or reality of such forces" [3] p. 69. This looks tame enough, since "phases one and two" are purely mathematical and so apparently don't concern themselves with existence problems.

In essence, the effort to segregate the mathematical backbone from its physical interpretation has been the standard gambit of the empiricist tradition, from Galileo to Mach, Duhem and Heisenberg. Its general idea is that the mathematical structure is hypothesis-free, whereas the physical interpretation is *merely* hypothetical. Those thinkers who strove to free science of its hypothetical luggage argued that the physical interpretation is not part of the essence of science, since it changes as science grows and is left behind as a useless garment. What remains unchanged throughout the development of science is its mathematical structure which thus becomes the essence of scientific procedure. This mode of saving the respectability of science in the face of its changeability and development is expressed in various instrumentalistic apologies, according to which the mathematical backbone of science is *autonomic* and does not commit the scientist to any existential commitments. These apologetics have developed in refinement since Osiander epitomised them in his cheat-preface to Copernicus *De Revolutionibus*, but the main change in their import has been that whereas Osiander defended Copernicus, Mach and Duhem defended science. The modern conception of hypothetico-deductive science is, again, geared to defend the respectability of science in much the same way: the physical interpretation, it says, is merely and always hypothetical, and so the scientist is never really committed to it. Hence, when science sheds the physical interpretation off its mathematical skeleton as time and refutation catch up with it, the scientist is not really caught in error, for he never was committed to this interpretation in the first place. This is the apologetic essence of present day, Popper-like, versions of the idea of science as a mathematical-core-cum-interpretational shell. This is also Cohen's view, for it aims to free Newton of any existential commitment to which his theory might allegedly commit him.

It will be readily seen that Cohen regards this methodological distinction between mathematics and physics to be the backbone of the Newtonian revolution in science (which is, in its turn, the climax of the whole Scientific Revolution) for a very clear reason: it enables us to argue that Newton could use freely the new concept of centripetal force, even though he did not believe in physical action at a distance and could not conceive how such a force could act to produce its effects". ([3] pp. 22–3) Thus, Cohen takes the climax of the Scientific Revolution to be a methodological rule which permits and allegedly enables the scientist to work without understanding ([3] p. 56) and in full disregard of the nature of the physical entities, which his mathematical symbols signify.

These points present difficulties which make it impossible for me to regard this model of Newton's science as anything more than a version of traditional apology which characterises recent philosophy of science. First, even though I am quite convinced that Newton himself would like to be regarded through Cohen's interpretation, as some of his famous pronouncements testify, he simply had another philosophy of science which was in flat contradiction with Cohen's Newtonian style, and which was infinitely more fundamental to him. Secondly, even if Cohen's interpretation may be ascribed to Newton, it suffers from internal difficulties which turn this ascription, if it be true, into a source of puzzles of such dimension as to make it implausible, to say the least.

Firstly, then, Newton's hatred of hypothetical science is basic to the structure he gave to his physics. His slogan, *Hypotheses non fingo*, is anything but a mere expression of misguided emotion. But neither does it represent a neutrality-policy, a non-commitment declaration. On the contrary, it expresses his most derisive attack on traditional ways of doing science, a tradition which actually viewed science as a hypothetico-deductive enterprise. Newton was engaged, from his first scientific publication onward, with an attack against this hypothetico-deductive philosophy, and his effort and frustration fed on a recurring failure to impress his own revolutionary method on his contemporaries. This central fact is very seldom appreciated today (and among philosophers and historians of philosophy he is regarded as a hypothetical scientist, as see [12], pp. 231–2), and needs to be emphasised in some detail.

Newton used to accuse Descartes, Leibniz "and some others" of basing their philosophies on hypothesis, and such systems he sometimes named "Hypothetical-Philosophy:"

> Hypothetical Philosophy consists in imaginary explications of things and imaginary arguments for or against such explications, or against the arguments of Experimental Philosophers formed upon Induction. ([8] V p. 397)

His main point is compounded of two subsidiary points: First, hypothetical philosophy is based on imaginary arguments aiming to explain things. Second, it is based on imaginary arguments against explanations given by other philosophers, and mainly by "Experimental Philosophers". Now, the first point is simply that the principles or basic axioms of Hypothetic Philosophy are not "derived" or "deduced" from phenomena, but are openly *declared to be assumed*, declared to be "hypotheses" in the literal sense. The reason for this was, of course, the fact that these axioms were, as a rule, concerned with the unobservable and invisible realm. What Newton argues is that this fact renders them *uncertain*, and hence *possibly wrong*. Since he intends this as a critique, it follows that he regarded Hypothetical Philosophy a wrong way of doing science simply because it could be false or in other words, because it was not certainly true. By contrast, Newton's way of doing science, what he uses to call "Experimental Philosophy" is exactly this: it is certainly true.

This connects with the second point of his accusation against Hypothetical Philosophy. The standard reaction of his critics was, usually, to deny his main claim, what he himself regarded as his main achievement, namely, the creation of a certain explanation of facts in contrast to the merely hypothetical explanation which existed to his day. The reaction was simply to deny this claim and to argue that Newton's explanation was in first just as hypothetical as any other around. To prove this, two independent methods were used.

One method was to point out where exactly Newton's principles were neither facts nor deduced from facts. This method was used by people who sometimes had nothing to do with Cartesianism, such as Hooke (in his criticism of Newton's *Optics*). This was also how Huygens and Leibniz criticised Newton's theory of gravitation, a criticism that was adopted also by Newton's desciple Cotes.

The second method was to prove the existence of an equivalent but different explanation. The idea was that this proves at once the uncertainty of the Newtonian explanation, for if another explanation is *possible* then none is *necessary*, and if it is not necessary it is also not certain. In sum, the existence of an equivalent theory proved the hypothetical status of both theories. This method was used, again, by Hooke in his criticism of Newton's optics, and by Leibniz in his criticism of Newton's gravitation theory. The equivalent theories which they used were, of course, their own.

This last circumstance led Newton sometimes to regard his opponents as though they were arguing for the truth of their own pet-theories and the falsity of his. The way he interpreted their attack, their argument was this: Since our theories are true, it follows that yours is false. And the truth of their theories was founded on metaphysical arguments. But, Newton retorted, this is nothing but romance and imagination, and hence it cannot be regarded as a proof. "Imaginary arguments" were, in his terms, metaphysical arguments which pretended to prove the truth of a theory by *imagining* what the system of nature must be like in accordance with metaphysical principles.

This was a mixup, of course. Leibniz, or Hooke, constructed an imaginary world in accordance with their metaphysics of matter. But this was not then used as a refutation of the truth or falsity of any other theory. Since the imaginary world was not patently false, it followed that Newton's theory was not necessarily true.

Newton's reaction was to point out the inferiority of the offered equivalents. And this just may be taken as an indication that he in fact understood exactly what was going on. I mean, since he answered by refuting the equivalence of the alternatives, he obviously saw that what was offered was an equivalent, and hence he also saw that the intended argument was not to prove the *truth* or falsity of any theory.

This is what usually triggers his attacks on ad-hocism. Obviously, he implies, there is no limit to the number of equivalents imaginable, if no limits are put on the method of inventing them. But it is exactly the fact that their invention is "formless" that proves their inferiority as compared with an invention that is informed by strict limits. Ad-hocism is just such a non restricted explanation, and Newton's argument is summed up by stating that this proves its hypothetical status.

So the point of his argument is that no equivalence is thereby proved, for his own theory is not hypothetical, whereas his opponents base their allegedly equivalent theories on hypothesis.

The situation now is this. Newton's opponents argue that his theory is hypothetical because they can produce equivalents; and Newton answers that these are not equivalents since they are hypothetical while his theory is certain.

He was wrong, of course. What he does is to make the equivalence argument hinge on some formal characteristic which is identical in the two equivalent theories. But why should this formal trait be accepted as a prerequisite for being an equivalent? After all, the equivalence which is aimed here is not more than vis-à-vis the *possibility* of being true and explanatory. Not, mind you, being *equally* possible and equally explanatory, but merely being *both possible* and *both explanatory*.

This is exactly where Newton usually misses the logic of the situation: if we demand equivalence as to all formal traits, there will be small chance of ever coming up with real examples of equivalents. This means that the equivalence argument in its capacity as disproving the necessity and certainty of a given theory would be impossible to apply.

His argument against ad-hocism, namely, that ad-hocism allows for an indefinite amount of liberty in constructing equivalence arguments is associated with his latent criterion of truth. And here it becomes clear that Newton knew all the time that his principles are neither phenomena nor derived from phenomena. For his criterion is the harmony criterion: if a small number of principles explains a large number of phenomena, then it is probably true. And if it explains all phenomena, then it is certainly true.

Obviously he saw that a criterion of truth was needed for his own theory, that this criterion was not derivation *from* phenomena but rather derivation *of* phenomena, and not just any derivation but only a mini-max one, namely, from minimal principles to maximal phenomena.

Though the proper method of science has been conceived, ever since Aristotle codified the idea, to be the twin process of analysis and synthesis, by the second half of the 17th century only the synthesis stage attained a solid, well formed pattern in those areas of physics (such as statics, geometrical optics, astronomy), that reached a mature mathematical form. In these highly sophisticated structures, theories were put forward starting with a set of assumptions, usually highly abstract and theoretical, and concluding with rigorously deduced consequences of concrete observationally testable nature. However, synthesis was by far the less important stage, for it was well known that no matter how rigorously these consequences were deduced and however precisely tested and verified, this did not elevate the theory one step towards the most highly valued goal of science, namely, *the certain knowledge of causes*. Synthesis was ideologically worthless, for the theory thus expounded would always remain merely probable. And while synthesis thus achieved the high technical perfection typical of the ideologically worthless, nothing was actually known about handling the ideological core of the procedure, namely, analysis or the derivation of causes from phenomena. When, in the 17th century the mechanistic conception of the world became the leading slogan in the new science, the despair from ever achieving the discovery of causes became even more acute, since these causes – the elementary atoms or corpuscles – were regarded and defined as invisible. This situation generated an official skeptical outlook and code of scientific behavior, which was most eloquently formulated by Descartes, Boyle, Huygens, Hooke and Glanvill.

Newton's 1672 announcement of his revolutionary optical theory, was put forward as a challenge to this official skepticism. Newton declared here, that he had succeeded in putting analysis, or the discovery of causes on a firm basis, as firm, at least, as was that of the traditional synthetic geometrical optics.

The physics itself was quite simple and there were no difficulties in understanding the new physical principles. What the critics, such as Hooke, Pardies and Huygens, found incomprehensible were the certainty claims which Newton insistently announced. These claims were regarded as scandalously unbalanced, and Oldenburg decided to omit, on the publication of Newton's 1672 Optical paper, a passage in which the then obscure Cambridge professor asserted the revolutionary implications of his discovery:

> A naturalist would scarce expect to see the science of those [colours] become mathematical, and yet I dare affirm that there is as much certainty in it as in any other part of optics. For what I shall tell concerning them is not an hypothesis but most rigid consequence, not conjectured by barely inferring 'tis thus because not otherwise, or because it satisfies all phenomena (the philosophers universal topic) but evinced by the mediation of experiments concluding directly and without any suspicion of doubt. [8] vol I p. 97.

To Hooke who, as a referee, has read the unexpurgated version, this was raving nonsense. No theory in physics could possibly be a "most rigid consequence" of experiments, "concluding directly and without any suspicion of doubt." Theories were never "necessary". So the greatest part of his review letter was an attack against this apparent dogmatic naïveté of the newcomer. Newton never succeeded in convincing Hooke, nor anybody else for that matter.

Then, in 1687, he published the second analysis argument, again fully mathematized, in which he deduced the law of inverse square gravitation from astronomic phenomena. Again, the leading scientists of Europe, such as Huygens and Leibniz, consented to regard the theory of gravitation as, at most a good hypothesis, and the 1672 scene was re-enacted. To Newton's mind the central thesis of the *Principia* was, again, as certain as the phenomena at its basis and as the mathematics which deduced it from them. Again he succeeded in accomplishing the unbelievable, namely the mathematization of another analysis argument, and so he declared in the preface:

I offer this work as the mathematical principles of philosophy, *for the whole burden of philosophy seems to consist in this – from the phenomena of motions to investigate the forces of nature*, and then from these forces to demonstrate the other phenomena: and to this end the general proposition in the first and second Books are directed. In the third Book I give an example of this in the explication of the System of the World; for by the propositions mathematically demonstrated in the former Books, in the third *I derive from the celestial phenomena the forces of gravity* with which bodies tend to the sun and the several planets. Then from these forces, by other propositions which are also mathematical, I deduce the motions of the planets, the comets, the moon, and the sea. [6] p. xvii.

In the next years Newton repeated this revolutionary message again and again, and with each new edition of the *Principia* and the *Opticks*, the success of the two mathematized analysis was transformed into a new methodology of natural science quite appropriately presented as the realization of an old and established philosophy according to which the natural philosopher was the discoverer of the true, unobservable reality hidden beneath the phenomena. The *Optics* of 1706 was concluded by a synoptic view of the work as a huge analysis–synthesis argument:

As in mathematics, so in natural philosophy, the investigation of difficult things by the method of analysis ought ever to precede the method of synthesis. This analysis consists in making experiments and observations. Thence we may reason from compounds to simple things and from motions to the forces producing them: and in general from effects to their causes. [7] p. 404.

And Cotes, who was the first Newtonian to publish a popular epitome of the *Principia*, constructed his whole account, in the preface to the second, 1713 edition of the *Principia*, around the new mathematized analysis, the process whereby

[Proponents of the "experimental philosophy"] derive the causes of all things from the most simple principles possible; but then they assume nothing as a principle, that is not proved by phenomena. They frame no hypothesis, nor receive them into philosophy otherwise than as questions whose truth may be disputed. They proceed therefore in twofold method, synthetical and analytical. From some select phenomena they deduce by analysis the forces of Nature and the more simple laws of forces; and from thence by synthesis show the constitution of the rest. This is that incomparably best way of philosophizing, which our renowned author most justly embraced in preference to the rest, and thought alone worthy to be cultivated and adorned by his excellent labors. [6] p. xx.

The ancient, hypothetico-deductive methodology was dead. A new era was established, the era of the methodology of the deduction of causes.

I have argued, till now, that the traditional view of Newtonian science, entitled by Cohen the "Newtonian style", is false. This view seeks to establish the neutrality of Newton vis-à-vis the physical interpretation of his mathematical theory, by arguing the separability of these two parts of his science from each other, and the feasibility of disregarding the nature of the physical entities which the mathematical signs denote. I have pointed out the closeness of this view with the hypothetico-deductive view, and then went on to argue that Newton rejected any hypotheticism, and that his main claim to have revolutionized science was exactly his claim to have established science on a certain, non-hypothetical, foundation. This he believed to have accomplished through his use of the double method of analysis and synthesis. In this method he *deduced* causes and principles from phenomena, which conferred upon the invisible causes and principles the certainty of facts.

There is, however, a more fundamental reason for rejecting the traditional view of Newton's science. Cohen's version of this view incorporates also the traditional belief in some kind of interaction going on between physical reality and the construction of the mathematical structure, an interaction which aims to "correct" the mathematical theory by comparing it with the physical world. As Cohen puts it, first a mathematical structure is built. Then it is compared with physical reality and corrected as needed. Then it is rebuilt, incorporating the new improvements dictated by reality. Then it is compared again with reality, and so on. Even Newton himself describes his work along some such lines, says Cohen.

But surely there is no evidence that Newton was telling the truth. In fact, an alternative model is much more likely to be the real one: He starts with a simple mathematical model, such as consisting of one mass-point moving in a force-field. He then turns it into a physical model: The force-field gets a source which is some mass, and the mass-point becomes an extended mass.

Since the laws of motion dictate that the central mass cannot be stationary *if it acts directly* on the revolving mass, and since Newton assumes it actually acts directly, it follows that the motion will not be strictly Keplerian. Hence Newton concludes that the motion of the planets is not strictly Keplerian. It is the physical theory which dictates this, not physical reality. Hence it is not reality which corrects the theory, but rather the theory which corrects reality.

There is hardly any sense in saying that replacing a mass-point by mass-balls, and placing another mass at the origin of a force-field, is a result of comparing the mathematical structure with reality. This is simply the transformation of a mathematical structure into a physical theory. And what comparison now follows is more of a dictation than a comparison: the physical theory tells us what it is that *observable* reality is, namely, it tells us that the planetary system cannot be strictly Keplerian.

But it tells this on condition, namely, that the force which acts on the planet is *directly* enacted by the sun. Now, Newton had some arguments for the *mutuality* of gravitational forces, but none for their *directness*, and these are not equivalent: in the case of pressure, for example, the piston and the walls of the cylinder act mutually on each other, via the enclosed gas, but since they don't act directly, the third law does not apply here. Directness, on the other hand, obviously implies mutuality as well as the application of the third law.

Since Newton had no argument whatsoever for the directness of gravitation, that means that the directness of gravitation was an assumption which his physical theory assumes. It is a hypothesis, and an unwarranted hypothesis at that. Newton constructed his physical theory pressupposing this hypothesis, and only via this presupposition it followed that the planetary motion was not Keplerian. Cotes pointed out this logical or structural fact to Newton, argued that this demolished Newton's claim to having achieved a certain, non-hypothetical theory, and was patronisingly but promptly and firmly answered by Newton. Cotes never dared raise this point again, though he had his own way of getting back at the master.

This was a capital point, for if directness was a mere hypothesis, then universal gravitation, being based on this hypothesis, was no more than a *presupposed* hypothesis as well. It is not proved by the theory, but is presupposed by it at the very first step of its construction. It is built into the theory at the moment Newton decided to make the force by which the central mass acts on the planetary mass a direct force. I say he decided, for he obviously could have chosen to make it non-direct, and hence to avoid the application of the third law, and hence to avoid the "derivation" of the universality of gravitation.

A major point may now be considered: how can we decide whether or not Newton was committed to the existence of directly-acting forces between masses? As Cohen rightly implies – the fact that a theory is constructed does not, by itself, signify any commitment as to the existence of its theoretical entities.

However, if the constructed theory is then proved to be true, this proof – whether valid or not – does most certainly express a commitment as to the existence of its theoretic entities. And Newton constructed such a proof, for this is exactly what the "analysis" was all about: from the fact that the planetary motion is Keplerian he deduces the existence of centripetal forces acting on the planets and having their origin in the sun. And then he deduces the existence of a reciprocal force, acting on the sun and originating in the planet, in accordance with the third law of motion. In Corollary I to Proposition V of Book III, Newton makes this explicit: "Since all attraction (by law III) is mutual, Jupiter will therefore gravitate towards all his own satellites . . . ."

So, he assumes that the third law of motion applies in the case of gravitational forces across spaces: Central bodies are here proved to be attracted towards their satellites since the satellites are assumed to react to the forces with which the central bodies act on them. Since Newton undertakes to prove these statements to be true, he obviously commits himself to belief in the existence of such things as planets, sun, and gravitational forces. He obviously also commits himself to belief in the direct action of one body on the other. In other words, he is committed to the "nature" of these forces, namely, that they are direct and act "immediately" at a distance. Moreover, to repeat a point, whether or not to apply the third law of motion is a problem which Newton never solved by logical derivation. He just decided what to do.

Now, the fact is that Newton is on record as denying he assumed direct action, (his letter to Bentley) or as being at least indifferent as to whether this action was direct or indirect. This action, he wrote, could arise either from the bodies "mutually seeking one another" (which is a direct action I suppose), or if they "agitate one another by means of spirits emitted", or "from the action of aether or of air or of any medium whatsoever". This indifference brings Cohen to assert that Newton "seems not to have believed in attraction as a primary quality." [3] p. 70.

Whether someone presupposes or is committed to something or not is not a matter of his declarations. He may be mistaken, or he may have reasons for concealing or misleading his listeners. It is, primarily, a matter for logical analysis, which may be followed by historical consideration of the bearing of this logical analysis on the whole historical material at hand. Surely we cannot draw any conclusions from Newtons' commitment-declarations to his real commitments: logically, these two are irrelevant to each other, and I don't suppose we are ready as yet to leave off the strictly logical sense of commitment.

This was essentially Cote's intention when he wrote to Newton that "till this objection be cleared, I would not undertake to answer anyone who should assert that you do *Hypothesim fingere*." ([4] p. 153) That is, what decides whether or not Newton builds on hypothesis is none of his declarations of belief, but exclusively, the logical structure of his theory. And this logical structure rather decisively shows, says Cotes, that Newton "seems tacitly to make this supposition that the Attractive force resides in the Central Body". (ibid.) In short, commitments are the statements, whether "tacitly" or explicitly made, of the theory when this theory is asserted to be true. Commitments are not actual beliefs, rather they are what one must believe on pain of being inconsistent. In this sense a theory is a statement of the necessary and actual concatenation of potential commitments, and hence of consistent beliefs. Hence, even at the purely mathematical stages of the work, when no truth declaration is yet made, some very definite commitments are already at work, namely commitments as to the necessary connections and implications of possible beliefs. Hence when Newton writes that he uses

> the words attraction, impulse, or propensity of any sort towards a center, promiscuously and indifferently one for another, considering those forces not physically, but mathematically: wherefore the reader is not to imagine that by these words I anywhere take upon myself to define the kind or manner of action . . . [6] p. 5–6.

he is telling a half truth, at best. Thus, once he uses the third law of motion to derive the necessary existence of a correlative force from a hypothetically assumed force, he is thereby committed to the "kind" and "manner of action" of these possible forces. In particular, he is thereby committed to their being directly acting forces. This is both "kind" and "manner of action", even though no simple existential commiuments are involved.

Let me now move from tacit, or logically entailed, commitments to declared and avowed commitments. These must be taken at their face value whenever they are the sole source of our knowledge and do not contradict the logically, entailed commitments of the theory. Such are Newton's declared commitments on the subject of God's place in his system of science.

The long-standing dispute between those scholars who insist on the metaphysical foundations of Newton's physics and those who regard his metaphysics as a mere emotional appendage to his science proper, is closely bound to this historiographic problem of commitments. Thus, a recent attempt to present Newton as the founder of 18th century's Deistic-positivistic attitude, uses the argument that since Newton held God's activity in this world to be knowable only a-posteriori, God's position in the world was changed, in Newton's mind, from that of law-giver to that of watch-maker. Hence the expression "laws of nature" signifies either the constants of a self-regulating watch, or the "purely human conventions concerning phenomena whose essence remained unknown." [2A] The fact that Newton used the double diction "*Axiomata seu leges motus*" shows this nominalistic attitude, where "law" is intended in "functional, non-substantive" sense and signifies not "God's eternal decrees" but merely the "categories of our experience". [2] p. 283.

This argument is obviously a nice example of the apologetic nature of some contemporary Newtonian Historiography. The effort to present Newton as a sober positivistic, no-nonsense scientist fit for incorporation into the venerable origins of the 20th century tradition leads scholars to attribute to him ideas about the autonomy of science which are strangely out of tune with his declared commitments.

Metaphysics is both a starting point and the final aim of Newton's physics, and the idea of autonomic science may be found in Newton only in a very limited sense. At most, one may argue that Newton's physics is autonomic in the logical sense, namely, all its theorems follow from its axioms. But as such, it is a meaningless system, and obviously not a physics. Meaning is injected into it only by basing it on an additional act, namely, the specification of the meanings of its basic terms. These meanings need separate specifications for a crucial reason: they cannot be read off the theory even when all its procedures are given. Newton quite deliberately distinguished between meanings and procedures, and underlined this basic distinction by insisting on the difference between the absolute and the relative, and parallelly, between the ultimate aim of his theory and its starting point. Whereas the starting point is the phenomenal and experimental, and hence the relative and procedural, the ultimate goal of the Newtonian science is the transphenomenal, the absolute basis of physical reality, in short, the metaphysical. This is, in essence, the raison d'être of Newton's analysis-synthesis methodology: as is well known, this twin method stems from Aristotle's doctrine that the last in the order of discovery is prior in the order of nature, (and correlatively, the first in the order of discovery, that is, the phenomenal is the last and the least from the absolute point of view of nature).

The evidence, given by physical phenomena, was to Newton quite univocal as to the causal relation of the world and of God. It proved to him that God is in fact an eternal and constant sustainer and could in no way be regarded as a watchmaker. S. Clarke's famous words to this effect leave hardly any doubt as to the final attitude of Newton:

> The notion of the world's being a great machine, going on without the assistance of a clockmaker is the notion of materialism and fate and tends (under the pretence of making God a supra-mundane intelligence) to exclude providence and God's government in reality out of the world.

And, of course, the evidence for this rejection of God the watchmaker comes to Newton from his physics. Near to the beginning of his *Principia* he lays the conceptual ground work for the protracted argument that constitutes the book, and he treats of the nature of space, time, mass and in an implicit way also of the logical and philosophical grounds of his mathematics. Twenty-seven years later he was persuaded that God's place in the world was a subject that could not be left out of his physical theory, and so he added to it his General Scholium. He was then seventy-one but it turns out that his metaphysical interest was life-long. The view of physics and metaphysics as one indivisible block was inevitable to him, for physics was the inquiry into the nature of bodies in space and time, but since space and time were the abode of God, there was no possible way to demonstrate separate departments of inquiry within all this.

The theological ultimate goal of Newton's physics is a key to a solution of what, to me, is the single most puzzling fact about the Scientific Revolution, namely, its invention of the infinitestimal calculus. The puzzle here is well known: what is it that the Greek mathematicians lacked which blocked their way to the calculus? They were well acquainted with the idea of a limit of an infinite converging series, of squaring curvilinear areas by proving them to be the limits of two converging series of inscribed and circumscribed polygons. And yet their representative figure is Zeno who demands an explanation for the fact that Achilles catches up with his tortoise. And he does not get it, for the basic rational datum is that an infinite series does not have a last term and cannot, therefore, reach the limit. Aristotle disdainfully refers to Antiphon the Sophist who does not understand this elementary basis of geometry for he believes that the inscribed polygon *will* eventually coalesce with the circle. Basically what operated here was the rejection, out of hand, of an existent actual infinite. Aristotle is explicit on this, reminding his audience that whenever the geometer uses the word infinite, all he actually means by it are always finite sections, large as you may wish, but finite all the same.

This logical gap between the finite and the infinite, a gap represented by the ever existing gap between any term of an infinite series and its limit, is reflected in the Greek cosmology in the necessary gap between god and the world, or between his ideas, that is, the blue-print of the world, and its actualisation in matter. The material world, Plato tell us, is not exact. Even the loftiest of the material realms, that of the stars, lacks exactness, and the

genuine astronomer . . . will think it absurd to believe that these visible material things go on forever without change or the slightest deviation, and to spend all his pains on trying to find exact truth in them. [10] p. 243 (529B)

The relation between the "laws" of astronomy and the plan of the god who "framed" the sky, is the same as the relation between the diagrams which may be "exquisitely drawn by some consummate artist like Daedelus" and the truths of geometry. The reason is that the god who framed the world is merely a very accomplished Daedelus, an artificer, and though "the sky with all that it contains has been framed by its artificer with the highest perfection of which such works are capable" (ibid), this is not absolute perfection. The reason, Plato explains somewhere else, is that this artificer god is not a creator but merely a framer: Matter, along with some of its basic properties, is eternal and so presents a real resistance to the artificer's attempt to "frame" it according to his ideas. Plato's famous injunction to the mathematicians to "save the phenomena" may have been intended as an attempt to discover the blue-print of the sky – the laws which matter never quite follow. The gap between the idea and its materialization is here a reflection of the gap between the finite or well-defined and the infinite or formless.

The Christian God, having created the world ex-nihilo, cannot be conceived except as an infinite power enforcing the world into a strict and exact obeying of exact laws. Moreover, the world of the 17th century science is, in fact, actually infinite, and the infinite God sits in it everywhere. The idea of an actual infinite, everywhere present and in touch with the finite, constantly reverberates through mathematics. Inscribed polygons start coalescing with their circles, and infinite series finally reach their limits, in a miraculous feat of non-logic. Newton could firmly declare that we have a positive – not merely a negative – understanding of actual infinity when we *compute* the areas and volumes of figures and solids of infinite length. And he adduced as an example of such a positive understanding the triangle with two right angles at the basis:

for if a triangle is actually drawn, its sides are always in fact directed towards some common point where both would meet if produced and therefore there is always such an actual point where the produced sides would meet, although it may be imagined to fall outside the limits of the physical universe. And so the line traced by all these points will be real though it extends beyond all distance. [9] p. 134.

And when Newton wrote that by his ultimate ratios of evanescent quantities he meant

the limits to which the ratios of quantifies, decreasing without end, always converge and to which they may approach nearer than any difference, but can never go beyond nor attain to. He could add, without batting an eyelid, unless the quantities are diminished infinitely (*in infinitum*) [6], p. 39).

And of polygons inscribed and circumscribed within a curvilinear figure he nonchalantly asserted that because their difference becomes "less than any given"

the figures inscribed and circumscribed become ultimately *equal* one to the other, and much more will the intermediate curvilinear figure be ultimately *equal* to either. (ibid., p. 30)

The *Principia* opens with a preface which does two things: it first differentiates between geometry and mechanics, and then announces the book as a vast analysis and synthesis program. The discussion of the relation between mechanics and geometry is, now-a-days, somewhat neglected for its intention is not very clear. Newton says that whereas geometry is "that part of universal Mechanics which accurately proposes and demonstrates the art of measuring," and likewise "rational mechanics ... (is) the science of motions . . . accurately proposed and demonstrated," practical mechanics is not "perfectly accurate" for it depends on manual art. Hence, it comes to pass, Newton goes on to say,

that what is perfectly accurate is called geometrical and what is less so is called mechanical. But the errors are not in the art, but in the artificer. He that works with less accuracy is an imperfect mechanic and if any could work with perfect accuracy he would be the most perfect mechanic of them all. [6] p. xvii.

Two concepts are assembled here under the single term "accurate" – the accurate as the logically rigorous, and the accurate as the non-approximate. Whereas the first "accurate" applies only to arguments, the second applies primarily to physical entities such as lines. Newton has in mind not only the rigorous proofs of geometry but also its ideal geometrical entities. So when he goes over to practical mechanics he thinks only about its non-ideal, material entities. These render practical mechanics inaccurate because they are themselves only approximate but this, however, does not touch the logical vigor of practical mechanics in his eyes. Universal mechanics comprises rational mechanics which deals with ideal entities, and practical mechanics which deals with physical ones. But both are rigorous to the same extent from the logical point of view.

And there is something more going for practical mechanics: There is nothing wrong with it as an art, for as such it is as perfect as geometry. Its only blemish depends on the degree of accomplishment of the artisan. But since the world's artisan is "the most perfect mechanic of them all," obviously this world is a work of "perfect accuracy," and so practical mechanics which deals with motions in this world is as accurate, both logically and ontologically, as rational mechanics.

Newton next goes on to point out the plan of the book as an analysis-synthesis structure, that is from motions to forces, and back from these forces to motions. But it is presented twice: first as rational mechanics dealing with ideal motions, and then, in book III, as practical mechanics. In this book he shows that since the planetary motions are observationally Keplerian, the force that causes them must be the inverse-square gravitation. However, if this force is universal, the synthesis stage must show that the motions cannot be Keplerian. And so the whole structure of inference is now in danger: either the motions are Keplerian but then the whole deduction is contradictory (for the synthesis ends with approximate Keplerian laws) or the motions are not Keplerian but then the analysis is wrong.

This is where Newton made his most important contribution to Empiricist Logic: he declared that even though Kepler's laws were false, yet they were almost true. Secondly, even though his whole inference procedure was contradictory, it was nevertheless almost valid. Newton's most important methodological heritage were the twin ideas of "almost true" and "almost valid." Here at last the Scientific Revolution came of age and the new empiricist philosophy came to be gradualistic in essence: degrees of truth and of validity were from now on the hallmark of empiricism.

The need for a gradualistic sapping of the bases of classical dichotomic logic is felt throughout the Scientific Revolution. It comes out forcefully whenever its leaders must admit that even though the laws of nature are strictly true and no deviation from them exist, yet it is a matter of experiential fact that none of their announced laws are strictly true. Galileo blames the air and the shapes of the ball and of the plane, and Descartes has his plenum theory of space to do the same job. Kepler, who knows exactly the measuring uncertainty of his ellipse, very quickly forgets it and regards his discoveries as unconditionally true.

This is also the essential feature of their ideology of certainty. They all must have certainty in their physics and they all declare that it is attainable, but in the same breath they also add that it is in fact not attainable. Facts tell them their physics is wrong but they declare it to be strictly true. Reason tells them their physics can be only probable but they declare it to be certain. Logic tells them that infinite series do not have a last term but they go ahead anyway and regard the limit as just such a last term.

This is what I propose to name the 17th century cheek principle. It is a pervasive psycho-methodological feature, which stems from and is constantly nourished by the presence of the infinite in closed homily spaces.

Of central importance here is the exact meaning of the concept of an exact law of nature. Within the empiricist framework, where measurement replaces naked objective magnitudes, there is only one way of conceiving and explicating the idea of an exact law of nature, namely, as a law which comes out true under an infinite series of measurements whose errors form an infinite series converging to zero. The exact law is the limit of this infinite series of measurements. This series must be infinite for the concept of measurement excludes the possibility of zero error. Thus the exact is now the limit of the approximate.

But now comes the rub. The limit is, in this case something fully existent and material. It is not a blue-print in the mind of a half impotent demiurgos, but the work of the omnipotent God "the most perfect artisan of them all" and so it is *perfectly* embodied in *our* world. Hence the measurement series must eventually arrive at it, behaving exactly as those "ratios of quantities" which approach to the ultimate ratio of their evanescent quantity and even "attain to it" if they are continued to infinity. And just as that "ghost of departed quantities," the instantaneous velocity, becomes by that process a full member in the community of velocities, so the exact law becomes an ordinary being, like any approximate law in its series, simply because the exact law is the last member of this infinite series. The net result of this maneuver was that the difference between the exact and the approximate law becomes now whittled away, as need be if rational mechanics is to be identified with practical mechanics, and if the inference in the analysis from the approximate laws of Kepler to the exact law of gravitation is to be valid.

Newton's final model of the logical inference in the *Principia* was, therefore, something along these lines: in *rational* mechanics we deduce *accurately* the law of gravitation from the *accurate* laws of Kepler; in the parallel *practical* mechanics we do the same, for even though we have only *approximate* laws of Kepler, yet since there is no categorical difference between the approximate and the exact the deduction here is *almost valid* and therefore as good as valid. Since we then deduce again these same *approximate* laws of Kepler from the law of universal gravitation, we have no contradiction in our whole chain of inference.

And so he could write to E. Halley, (20 June, 1686) that just as Kepler

> only guessed the planetary oval to be an ellipse, so Hook could know no more but that the proportion [of gravity] was duplicate quam proxime and only guessed it to be so accurately . . . There is so strong an objection against the accurateness of this proportion, that without my demonstrations . . . it cannot be believed by a judicious philosopher to *be anywhere accurate.*

And so he could treat the planetary orbits as circles concentric with the sun whenever he wished to (as in the 3rd book of the *Principia* and in drafts of the Scholium Generale), and hope to put through the idea that the fixed stars are so far from us and each other as to exert no influence at all. This is how his idea of a perturbation-science may be interpreted. Newton wished, I suggest, to show by his qualitative perturbation-theory in the *Principia*, that even though gravitation is universal, the planetary orbits may be Keplerian not merely observationally but rather accurately. One passage is especially striking, asserting that even though this may happen only when the planets are points devoid of mass, yet, in the case of actual masses, as these decrease, the perturbations

> may become less than any assignable . . . and . . . the orbits may become ellipses and the areas answer to the times, without any error that is not less than any assignable, [6] p. 171–2.

which, of course, are the exact and accurate Keplerian Laws.

University of Haifa

ACKNOWLEDGEMENT

The passages quoted from Isaac Newton, *Principia* (ed. by F. Cajori), are copyright 1934 by the Regents of the University of California and are reprinted by permission of the University of California Press.

REFERENCES

[1] Burtt, E. W., *The Metaphysical Foundations of Modern Science* (N.Y. 1954).

[2] Cassini, P., 'Newton, le lois de la nature et la "Grand Ocean de la Verite"', in *Proceedings of the XVth International Congress of the History of Science* (Edinburgh University Press, 1978).

[2A] Cassini, P., English abstract of [2].

[3] Cohen, I. B., 'The Principia, Universal Gravitation, and the "Newtonian Style"' this volume.

[4] Eddelston, J., *Correspondence of Sir Isaac Newton and Professor Cotes* (London, 1850).

[5] Hesse, M., 'Reasons and Evaluation in the History of Science', in R. Young (ed.), *Changing Perspectives in the History of Science* (London, 1973).

[6] Newton, I., *Principia* (Translated by Motte, University of California Press, 1962).

[7] Newton, I., *Opticks* (Dover, N.Y. 1952).

[8] Newton, I., *The Correspondence* (Cambridge, 1959 on).

[9] Newton, I., 'De Gravitatione etc.', in A. R. Hall and M. Boas Hall (eds.), *Unpublished Scientific Papers of Isaac Newton* (Cambridge, 1962).

[10] Plato, *The Republic* (translated by F. M. Cornford, Oxford, 1941).

[11] Strong, E. W., *Procedures and Metaphysics* (Berkeley, California, 1936).

[12] Rogers, G. A. J., 'The System of Locke and Newton', this volume.

I. BERNARD COHEN

THE *PRINCIPIA*, UNIVERSAL GRAVITATION, AND THE "NEWTONIAN STYLE", IN RELATION TO THE NEWTONIAN REVOLUTION IN SCIENCE:

Notes on the Occasion of the 250th Anniversary of Newton's Death

CONTENTS

1. INTRODUCTION*

Scholars generally agree that Newton's *Principia* marks the climax of the Scientific Revolution.[1] This great treatise epitomized the application of mathematics to natural philosophy and set a standard which scientists in such diverse areas as palaeontology and physical chemistry saw as the goal for all developing sciences.[2] Yet, to judge from the current literature in intellectual history, the history of philosophy, and even the general history of science, and despite the unanimity on the importance of Newton's *Principia*, there is a rather wide-spread lack of precision and a considerable variation from one author to another regarding the nature of the book – its contents, its method, its positive accomplishments, and its failures. Very few scholars today, save for Newtonian specialists, have read the *Principia* through from cover to cover – a task that demands mathematical skills beyond the range of most historians, plus a background in various aspects of physics and of celestial mechanics.[3] Newton himself was aware of the fact that to study carefully every proposition of the first two books of the *Principia* 'might be too time-consuming even for readers who are learned in mathematics', and so he said that he was 'unwilling to advise anyone to study every one of these propositions'.[4]

Z. Bechler (ed.), Contemporary Newtonian Research, 21–108.

Even those scholars who do read in the *Principia* are apt to limit their study to a few selected parts: usually the preliminary prose (the Prefaces,[5] and the Definitions and the Laws), the first three sections of Book 1,[6] lemma 2 of Book 2 on the method of fluxions and the conclusion of Book 2 in which Newton allegedly proves that the Cartesian vortices yield results that are inconsistent with the phenomenological evidence. Even Book 3, on the 'System of the World',[7] will not get a complete reading, although attention will be paid to the opening 'Phenomena' and 'Rules for the Study of Natural Philosophy',[8] the first 12 or 13 propositions, and the concluding General Scholium. This procedure robs the scholar of any possibility of appreciating the power of the Newtonian methods and the full scope of his achievement. In particular it completely inhibits any understanding of the Newtonian distinction between the domain of mathematics and the domain of physical reality, which I find to be the characteristic feature of the "Newtonian style".[9] Can it truly be said that the science of the *Principia* has been comprehended by anyone who has not fully grasped the significance for Newton of 'mathematical principles' and 'natural philosophy', the two components of the title of his magnum opus? The generality of Newtonian celestial and terrestrial dynamics cannot be understood without prop. 41 (sect. 8) of Book 1, 'Supposing a centripetal force of any kind and granting the quadrature of curvilinear figures, it is required to find both the trajectories in which bodies will move and the times of their motions in the trajectories so found'.[10] Such a limited reading of the *Principia* also rules out prop. 66 of Book 1, in which Newton attacks the problem of the motions of three bodies, each attracting the others 'with forces that vary inversely as the square of the distances', under certain conditions. In this proposition, and its set of 22 corollaries, Newton may be seen developing some of 'the chief effects of the disturbing action of a body like the sun on the motion of a body like the moon'.[11]

It will be seen below that a close reading of the *Principia* and the versions of the preliminary tract *De Motu* (written by Newton in the autumn or early winter of 1684) enables us to specify the essential step toward the formulation of the concept and law of universal gravity. Furthermore, the analysis of this step will show the crucial rôle of the third law of motion. In displaying the stages of development of Newton's dynamics leading up to universal gravity, I shall indicate the importance of what I have called the "Newtonian style", a blend of mathematical and physical thinking which was a necessary condition for the production of the *Principia*. This Newtonian style is the key to the Newtonian revolution in the exact sciences; it enabled Newton to use freely the new concept of centripetal force, even though he did not believe in physical action at a distance and could not conceive how such a force could act to produce its effects without the mediation of something else.

Newton himself was clear about levels of mathematical and physical discourse; the distinction between them and their relations to each other constitute, in fact, the essential ingredient of the Newtonian style. And yet the most acute students of Newton's thought seem either not to have believed what Newton wrote on these topics or to have believed that Newton did not mean what he said (or did not say what he believed).

The clarification of Newton's method in the *Principia* is directly related to a number of issues in the history of scientific thought. Among them are: the alleged rôle of Hermeticism in the formulation of Newton's concept of universal gravity, the possible re-orientation of Newton's philosophy of nature in about 1679, Newton's concept of force (especially forces acting at a distance), the difference between Newton's system of celestial dynamics and the end-result of his studies of optics and the theory of matter, and finally the possible reason why there was no Newtonian revolution in the biological or life sciences in the seventeenth and eighteenth centuries.

2. SOME SOURCES OF INACCURACIES AND MISUNDERSTANDINGS CONCERNING THE *PRINCIPIA*

There are a number of reasons why the literature concerning Newton's *Principia* contains many kinds of inaccuracies and misunderstandings. Some of these apply equally to all aspects of Newtonian scholarship, but others seem to be especially characteristic of current views concerning the *Principia*. Here there are errors and uncertainties concerning Newton's method of discovery and the dates when he made certain crucial discoveries. First and foremost among many sources of error concerning Newton's procedures and the dates of his major discoveries are his own statements, often written to support a particular chronology or mode of discovery that would establish his priority, or to defend a position which he wished to maintain in controversy. Here are a few samples. One of Newton's most famous and most often quoted autobiographical sketches, a draft of a communication to Des Maizeaux (written *c.* 1718), describes his discoveries in 1665 and 1666:

> And [in] the same year I began to think of gravity extending to the orb of the Moon, and having found out how to estimate the force with which [a] globe revolving within a sphere presses the surface of the sphere: from Kepler's Rule of the periodical times of the Planets being in a sesquialterate proportion of their distances from the centers of their Orbs, I deduced that the forces which keep the Planets in their Orbs must [be]

reciprocally as the squares of their distances from the centers about which they revolve: & thereby compared the force requisite to keep the Moon in her Orb with the force of gravity at the surface of the earth, and found them [to] answer pretty nearly. All this was in the two plague years of 1665 & 1666. For in those days I was in the prime of my age for invention & minded Mathematicks & Philosophy more then at any time since.[1]

This alleged failure to get a better agreement than 'pretty nearly' has given rise to considerable speculation about a supposed 'twenty years' delay in publication.[2] But the quoted extract presents a somewhat fictionalized chronology, a kind of "scenario" written out by Newton early in the second decade of the eighteenth century, approximately half a century after the events had allegedly occurred. The document reads as if Newton had made a comparison of the CENTRIPETAL[3] *force* keeping the moon in orbit and the force of gravity. But, in fact, in those days Newton had not yet developed a clear concept of centrifugal "force", much less any concept at all of a "centripetal force",[4] but had in mind a vague concept which he then called 'conatus recedendi a centro' or 'endeavor to recede from the center', which he had taken over from Descartes.[5] He had at this time no idea of a balance of *forces*, and no concept of an inverse-square *force*, and hence certainly no inkling of a *force* of *universal* gravitation.[6] In fact, the foregoing memorandum was written as part of Newton's attempted documentation of his discoveries in relation to the dispute with Leibniz on the matter of discovery (or invention) of the calculus.[7] Furthermore, Newton himself has cancelled this document by drawing a line through it from top to bottom.[8]

In certain other of Newton's MS memoranda, and in an anonymous article published by Newton in the *Philosophical Transactions*, there is a second example of how misleading a statement of Newton's can be. It is his assertion that he 'found out most of the Propositions in his *Principia Philosophiae*' by 'the help of the new *Analysis*'; 'but because the Ancients for making things certain admitted nothing into Geometry before it was demonstrated synthetically', Newton 'demonstrated the Propositions synthetically, that the Systeme of the Heavens might be founded upon good Geometry'.[9] As we shall see below, there is very little reason to suppose that the propositions in the *Principia* were discovered in any other way than that in which they were presented.[10]

A third and possibly even more striking example of Newton's attempts to mislead his readers occurs in his presentation of the lunar theory in Book 3 of the *Principia*. This example differs from the foregoing ones in that here the allegation did not remain in MS and was not cancelled, nor was it printed anonymously apart from the *Principia*, but was actually published in both the second and third editions of the *Principia*. In the second edition (1713) Newton introduced a new scholium following prop. 35 of Book 3, beginning: 'I wished to show by these computations of the lunar motions that the lunar motions can be computed from their physical causes by the theory of gravity'. Thrice more during the scholium he asserted: 'By the theory of gravity it was also established . . .', 'By the same theory of gravity . . .', 'By the same theory of gravity . . .'. The impression he was giving was that he had actually computed the inequalities in the moon's motion by applying gravitational celestial mechanics. In fact, it was one of the glories of the *Principia* to have begun to compute the motions of the moon from a physical theory (i.e., gravitational forces of perturbation), whereas up till that time it had been the mode to compute lunar positions from pure geometric configurations (combinations of circles, much in the manner traditionally associated with Ptolemy's *Almagest*). Hence the history of lunar theories changes abruptly with the publication of Newton's *Principia.*[11] This fantastic achievement, however, was far from complete and in fact Newton himself had to have recourse to the old ways, since his program of deducing *all* aspects of the moon's motion from physical causes had been a failure.[12] He gave the impression, however, that he had been more successful than his results warranted, and many uncritical readers have taken Newton's word for the deed ever since a book reviewer, writing about the second edition of the *Principia* in the *Acta Eruditorum*, stressed the agreement between the observed and predicted motions of the moon as proof of the validity of the theory of gravity and of Newton's 'divine force of intellect' and 'outstanding sagacity'.[13]

A second cause of inaccuracy in relation to the *Principia* arises from the conscious or unconscious attempt by scholars to foist upon Newton a particular philosophical point of view, such as to have Newton appear to have been a kind of Baconian empiricist or inductivist, or an early positivist, and so on. Here again, it is possible to find Newtonian texts – published and unpublished – which, if read out of context, seem to support one or another philosophical position. This is especially the case with respect to Newton and hypotheses, the rôle of induction, and analysis and synthesis. Closely associated with this source of misunderstanding is a third one: the natural tendency of scholars to try to make a single constant entity out of the changing and varied complexities of mind of this extraordinary man.[14] This tendency toward unification is revealed in the suggestion that Newton's most positive science may have originated in some kind of Hermeticism; in a most extreme example, not only the *Principia* but even Newton's work in mathematics would become the offspring of alchemy or a parallel twin-creation of the Hermetic art.[15]

The fourth of the major sources of error and misunderstanding is the effect of scholarly sampling, or of reading the odd sentence or statement out of context, without giving adequate consideration to the paragraph or treatise (or section of the treatise) in which it is embedded. This may be illustrated by Newton's famous slogan, 'Hypotheses non fingo' (or, 'I frame [*or* feign] no hypotheses').[16] This famous sentence has often been read as a general description of the Newtonian natural philosophy. On the basis of this statement alone, Ernst Mach welcomed Newton into the ranks of positivist philosophers.[17] I shall not enter here into yet another discussion of the meaning of the words in this slogan, but I would point out that their apparent generality may be seen to be somewhat restricted by their context, being placed in Book 3 of the *Principia*, in the General Scholium which serves as a conclusion in the second and third editions.[18] For in Book 3 of the *Principia*, in the editions in which this General Scholium appears, there are in fact two numbered 'hypotheses', and in the preceding Book 2 there is a single one – plainly labeled 'hypothesis'.[19]

Fifth and last, a source of misunderstanding about the *Principia* has been the assumption by historians and philosophers of science that Newton developed his special versions of such concepts as force, mass, particle, gravity, momentum, and universal gravity primarily by merely thinking about them in a purely philosophical way, that is, by a method of conceptual analysis in the manner of a philosopher of science, by a kind of abstract process of intellection or pure thought. As I see it, Newton's mature concepts and methods in the exact sciences did not primarily develop in his mind as abstract solutions to philosophical problems of motion,[20] but were produced most often during his attempts to solve *specific* problems of physics or mathematics. The latter include motion under certain specified conditions such as Kepler's laws, the dynamics of impact, the effects of the third law of motion, aspects of celestial mechanics such as perturbations. Hence the most fruitful way to gain an understanding of the formation and growth of Newton's concepts in dynamics is to examine in detail the Newtonian attack on particular problems: the kind of force that produces the area law, the Newtonian measure of force developed in section 2 of Book 1 of the *Principia*, the stages from central forces to mutual forces and from mutual forces to universal gravitational forces. There is more to learn about the genesis and transformation of Newton's scientific ideas by this method than by speculating on the possible origins of the concept of universal gravitation in the dark reaches of Hermeticism, or in Newton's psychological state of mind arising from the trauma in his never having known his father.[21]

In displaying some of these shared misunderstandings, my aim has been primarily to indicate certain paths for needed future research with respect to the *Principia.*[22] I see a quadrivium of needs, as follows. *First*, a better understanding of Newton's predecessors, and the background, sources, and development of his ideas. *Second*, a knowledge of the *Principia*, based upon its contents: the development of the subject matter, theorem by theorem. *Third*, an inquiry into the varieties of possible influences[23] of the *Principia* on those who created a new science in the Age of Reason, reaching a culmination in the *Mécanique céleste* of Laplace.[24] *Fourth*, there are all the questions of the lasting influence of the science of the *Principia*, and the fate and condition of Newtonian Science in an age of Relativity and Quarks.

There are, of course, further research topics dependent on these. Among them is the study of the influence of the *Principia* in realms outside of mathematical physics and rational dynamics and celestial mechanics. This influence had at least two forms. One: on other sciences which tended to look on the *Principia* as the goal to which all true science ought to tend.[25] Two: on the thinking in non-scientific areas such as political and social theory or practice, the behavioral sciences, and philosophy and theology. Additionally, there is the matter of the "Newtonian style", a way of doing science successfully which was used (but not set forth as a set of explicit principles or precepts) throughout the *Principia* and which, as I have said above, is of particular importance in the degree to which it constituted the essential feature of the Newtonian Revolution in the sciences.[26]

3. NEWTON'S ANALYSIS OF THE MOTION OF THE MOON IN THE 1660'S

Thanks to a century or more of textual and analytical studies – going back at least to those of S. P. Rigaud, J. Edleston, and W. W. Rouse Ball – we now know the main stages of the development of Newton's concepts and methods in dynamics. Among the scholars who have contributed to our present understanding are Eric J. Aiton, René Dugas, A. Rupert and Marie Boas Hall, John W. Herivel, Emile Jouguet, Alexandre Koyré, Aleksei Nikolaevich Krylov, J. A. Lohne, Ernst Mach, J. E. McGuire, Ferd. Rosenberger, H. W. Turnbull, R. S. Westfall, D. T. Whiteside, and many others.[1] There is no real unanimity concerning every question about Newton's dynamics and celestial mechanics, but there is near agreement on certain fundamentals. For example, we may be reasonably certain that Galileo was not the major influence on the development of Newton's dynamics, in the sense that Descartes was.[2]

Studies on the stages of Newton's dynamics have been characterized by an ever-increasing precision, the dispelling of mysteries and the clarification of ambiguities, and the rejection of established myths or fables. An example of this last category, to which reference has already been made, is that of Newton finding the inverse-square law of universal gravity in the 1660's and then witholding publication of it for two decades.[3] The story of Newton and the apple, however, although rejected by critics as a later or post-Newtonian invention (possibly by Voltaire), now turns out to be authentic, at least to the extent that it appears to have originated with Newton himself.[4] But this anecdote was told by Newton as part of his effort to have the discovery of universal gravity date back to the 1660's.

An outstanding example of a puzzle has long been Newton's claim that he had combined Kepler's third law with the rule for centrifugal "conatus" or "endeavour" in the 1660's. How could he have done so, since Huygens did not publish this rule until 1673, in his *Horologium Oscillatorium*? The factual details of this puzzle were charted by John Herivel, who found, interpreted, and published Newton's own derivation of this rule, dating from the 1660's. Newton's approach to the v^2/r rule was very different from Huygens's,[5] and is related to a paragraph in the scholium to prop. 4 in Book 1 of the *Principia.*[6]

It was thus perfectly possible that Newton, as he later claimed,[7] could have combined the Keplerian third law (r^3/T^2 = const.) with the centrifugal rule (v^2/r) to get to an inverse-square law.[8] All that he would have needed was the simplest algebra. On the assumption that planets move uniformly in circular orbits of radius r with speed v, having a period T of revolution, it follows that

$$\frac{v^2}{r} = \left(\frac{2\pi r}{T}\right)^2 \Big/ r = 4\pi^2\left(\frac{r^2}{T^2}\right)\left(\frac{1}{r}\right) = 4\pi^2\left(\frac{r^3}{T^2}\right)\left(\frac{1}{r^2}\right).$$

Since (by Kepler's third law) r^3/T^2 is a constant for the solar system, so that $4\pi^2(r^3/T^2)$ is a constant, it follows that v^2/r is proportional to $1/r^2$. Another way[9] of seeing how v^2/r may be combined with Kepler's third law ($T^2 \propto r^3$) to give $1/r^2$ is to start with $v = 2\pi r/T \propto r/T$, so that

$$v^2 \propto r^2/T^2 \propto r^2/r^3 \propto 1/r$$

whence it follows at once that

$$f \propto v^2/r \propto 1/r^2.$$

In Newton's earliest formulation of this law (in the 'Waste Book'),[10] he does not express his result in terms of speed and radius, but rather measures the force by its accelerating effect in linear motion. As paraphrased by Herivel, and here restated algebraically, Newton's measure of the force of a body's endeavour of receding from the center of a circle of radius r on which it moves with uniform speed v is as follows. This force has the same magnitude as a constant force (like gravity) which, during a time t, acts on a second (and equal) body so as to give it a speed v in a straight line, where t is the time for the first body to revolve through an arc r of length equal to the radius of the circle and v is the speed of the revolving body (thus $t = r/v$). This linear force is measured by the acceleration v/t. Since $t = r/v$, it follows that $v/t = v/(r/v) = v^2/r$. This Newtonian measure of centrifugal endeavour thus turns out to be equivalent to the familiar one, v^2/r.

Later on, Newton found the familiar measure of the centrifugal endeavour with which a body moving inside a circle continuously impels or presses the circle. It is, he said, as 'the square of the arc described in a given time divided by the radius'.[11] In uniform circular motion the distance traveled in any given time is a direct measure of the speed, and hence the centrifugal pressure is now merely said to be proportional to v^2 divided by r. By the late 1660's, as we shall see in a moment, Newton was not only measuring a body's centrifugal "endeavour" by the distance through which a second body would move under uniform linear acceleration, as in freely falling to earth, under a constant 'endeavour of approach' to the center which is equal in magnitude to the first body's 'endeavour of receding from the center'; he saw that this is equivalent to stating that D/T^2 or $D \times n^2$ is a quantitative measure of the 'endeavour of receding from the center', where D is the diameter of the circle, T the period, and n the number of complete revolutions in any unit of time (i.e., the frequency).

There is documentary evidence (see page 31) that Newton concluded in the 1660's that the planetary forces are as v^2/r and so proportional to $1/r^2$. If this is so for the planets, then why should it not be the same with respect to the moon? This line of thought *could have* led to a test of the moon's force by making a comparison of the distance fallen in 1 sec. toward the earth's center by the moon in its orbital motion and by a freely falling body on earth. And this series of steps is just what Newton later alleged he had done, to find that the two results (for the moon and a terrestrial freely falling object) 'answer pretty nearly'. That is, in the autobiographical statement written out for Des Maizeaux (and which has been discussed in §2 *supra*)

Newton said (*a*) that having found out how to estimate the force with which a globe presses the surface of a sphere, he used Kepler's third law to deduce 'that the forces which keep the Planets in their Orbs must [be] reciprocally as the squares of their distances from the centers about which they revolve', and (*b*) that he 'thereby compared the force requisite to keep the Moon in her Orb with the force of gravity at the surface of the earth, and found them [to] answer pretty nearly'. Now there is a document of the late 1660's (published and analyzed by A. R. Hall, H. W. Turnbull, and John Herivel[12]) in which Newton makes calculations which at first glance may appear to be somewhat like those mentioned in the later autobiographical memorandum. An analysis of this early document therefore allows us both to verify the accuracy of Newton's later assertions and to see how he conceived celestial forces and celestial motions in the later 1660's.

In this early document, Newton begins with a general computing rule for uniform circular motion: 'in different circles the endeavours [of receding] from the centers are as the diameters divided by the squares of the times of revolution, or as the diameters multiplied into the [squares of the] number of revolutions made in any given time'. This rule gives a result equivalent to the v^2/r centrifugal force rule. It says that the 'endeavour from the center' is proportional to D/T^2 or to $D \times n^2$ where D is the diameter, T the period, and n the number of revolutions per second (or per any other unit of time); but $v^2 = \pi^2 D^2/T^2$, so that $v^2/r = (\pi^2 D^2/T^2)/\frac{1}{2}D = 2\pi^2 D/T^2$ or $v^2/r \propto D/T^2$.

It is to be noted, however, that in the 1660's this rule is not computed for CENTRIFUGAL *force*, but rather for what Newton still calls a 'conatus recedendi a centro', an 'endeavour of receding from the center', a Cartesian concept which is far removed from Newton's later idea of a dynamical centripetal *force*. For a body at the surface of the earth, Newton uses his D/T^2 rule to determine the magnitudes of the 'endeavour of receding from the center' caused by the earth's daily rotation. He finds that on the equator, an 'endeavour of approaching the center [of the earth] by virtue of gravity' equal in magnitude to the 'endeavour of receding from the center' would impart a uniform acceleration to a heavy body which would cause it to move rectilinearly, starting from rest, through $\frac{5}{9}$ inch in 1 second. He then compares this quantity with the distance through which 'the force of gravity [*vis gravitatis*] is great enough to drive bodies downwards' at the earth's surface, which he computes to be 'about 16 feet in 1 second' or 'about 350 times farther in the same time than the Moon's endeavour [of receding] from the center'.[13] The next step is to use the D/T^2 rule to compare the moon's endeavour to recede from the earth's center with the endeavour of a body to recede if it is located at the surface of our rotating earth. He finds that 'the endeavour of the earth's surface at the equator is about $12\frac{1}{2}$ times greater than the Moon's endeavour of receding from the earth's center'. This leads him to two conclusions. The first is that 'the force of gravity is 4000 and more times greater than the Moon's endeavour of receding from the earth's center', a conclusion which follows from multiplying $350 \times 12\frac{1}{2}$, giving a result which is significantly more than 4000 (in fact, approximately 4400).[14] The second conclusion is that 'if this endeavour [of receding] from the earth results in the Moon's always turning the same face to the earth', then it must follow that 'the endeavour of this system of moon and earth to recede from the sun' must be less than 'the Moon's endeavour of receding from the Earth', since otherwise the moon would be constantly facing the sun rather than the earth.

This leads to the next assignment, to use the D/T^2 or $D \times n^2$ rule to compare the moon's 'conatus . . . recedendi a terra ad conatum terrae recedendi a sole' (that is, to compare the moon's 'endeavour of receding from the earth to the earth's endeavour of receding from the sun'), which turns out to be about 5 to 4, corresponding to a computed value of $559\frac{1}{2}$ to 100,000 for the minimum ratio of the moon's greatest distance from the earth to the earth's greatest distance from the sun. (These numbers and the argument about why 'the same side of the Moon is ever towards the earth' occur also in a letter from Newton to Oldenburg intended to be forwarded (at least in part) to Huygens in 1673[15]; here Newton uses the value 56/10,000.)

How does the foregoing analysis correspond to the assertions I have just called "(a)" and "(b)" from the later autobiographical memorandum? Newton (a) said he combined his rule for estimating the force with which a globe presses the surface of a sphere with Kepler's third law and deduced that 'the forces which keep the Planets in their Orbs' must vary inversely as the square of the distance. In fact, following the discussions of the endeavour of receding at the earth's surface, and of the moon's 'endeavour of receding from the earth' and the earth's 'endeavour of receding from the sun', Newton does calculate – in a supplementary paragraph – that 'for the primary planets, the endeavours of receding from the sun will be reciprocally proportional to the square of their distances from the sun', since 'the cubes of their distances from the sun are reciprocally proportional to the squares of their numbers of revolutions in a given time'. The difference between the early and the later documents is a profound one, however, since in the 1660's Newton is computing the planetary 'endeavours of receding from the sun', whereas he later said he had computed 'the forces which keep the Planets in their Orbs'. One is an individual tendency of each planet to fly away, but the other is a force exerted by the sun on every planet (and on comets too; in fact, on any body in the solar system).[16]

In statement (b), Newton later alleged that (i) he 'thereby compared the force requisite to keep the Moon in her Orb with the force of gravity at the surface of the earth' and (ii) found them to agree 'pretty nearly'. With respect to (i), even apart from the fact that Newton was discussing the moon's 'endeavour of receding from the earth's center' rather than a 'force requisite to keep the Moon in her Orb', the document of the 1660's contains no such comparison based on the inverse-square law he had deduced by using Kepler's third law. In fact, in that supplementary paragraph, Newton refers explicitly to the 'primary Planets' ('in Planetis primariis'), thus ruling out any implied application to "secondary planets" or planetary satellites. Nor is there so much as a hint that this restricted inverse-square law might possibly be generalized so that it need not be limited to only the "primary" planets and the sun but could be extended to the earth and moon and presumably also to Jupiter and Jupiter's satellite system. In fact, the *only* use to which Newton has put the inverse-square law in this document of the late 1660's is to tabulate the relative 'conatus recedendi' or 'endeavour of receding' at each of the planetary orbits, giving the sequence 614, 173, 91, 39, $3\frac{1}{3}$, 1. Had Newton tested the inverse-square law (even for 'conatus recedendi a centro', if not for "force" – and *certainly not* for "centripetal force") in the case of the moon, he would hardly have got the result which he described by the phrase (ii) agree 'pretty nearly'. He knew the radius of the moon's orbit to be 59 or 60 times the earth's semi-diameter; and 4400 could not have been 'pretty nearly' 3460 or 3600 but would have been off by about 22 per cent.[17] But the important thing about the manuscript of the late 1660's is that once Newton has made his calculations, he says *nothing whatever* about the significance of his result of 4400 (or, "4000 and more"). That is, he does not say that he has made a test and the result is a failure, wide of the mark; he does not say that the result implies that the inverse-square relationship is not valid when applied to the moon. For that matter, if the result had come out near to 3600, he would certainly have been puzzled, since the moon definitely does *not* move in a circle, as he well knew, nor even in an orbit that is even approximately circular. In fact, the moon's orbit has an elliptical eccentricity of $\frac{1}{18}$, which can hardly be neglected.

There is one final point to be made about the inverse-square law of 'conatus recedendi a centro' for the primary planets. In the document of the 1660's Newton did not use the rule of v^2/r but rather D/T^2 or $D \times n^2$. The algebra for combining v^2/r with Kepler's third law in order to get the inverse-square relation involves several steps, of which one – not to cancel out the r's in $4\pi^2(r^2/T^2)(1/r)$ – would not be at once obvious to everyone. But if the rule is expressed by $D \times n^2$ (as Newton does in the document of the 1660's), and Kepler's law is given in the form that D^3 is inversely proportional to n^2, it is a simple and immediately obvious step to replace n^2 by $1/D^3$ (for suitable units) so that

$$\text{conatus} = K_1 \times D \times n^2$$

$$D^3 = \frac{K_2}{n^2} \quad \text{or } n^2 = \frac{K_2}{D^3}$$

with the immediately obvious result that

$$\text{conatus} = K_1 K_2 \times \frac{D}{D^3} = K\frac{1}{D^2}$$

where K_1, K_2, and $K = K_1 K_2$ are constants of proportionality. It is no wonder that Newton believed that once the law of centrifugal force was known (as it was, generally, after being published by Huygens in 1673 in his *Horologium Oscillatorium*), very little credit needed be given for discovering that the force must then – for uniform circular motion – be inversely proportional to the square of the distance.[18]

In relation to Newton's testing of the moon's force and the force of gravity, there has been for a long, long time a certain amount of puzzlement arising from a statement made by William Whiston, based – as he said – on conversations with Newton 'soon after 1694'.[19] Whiston, like Henry Pemberton (who also said that he got the story directly from Newton himself[20]), mentions that in Newton's calculation (or "moon test"), he was, in Whiston's words, in some 'Degree, disappointed', since 'the Power that restrained the Moon in her Orbit . . . appeared not to be quite the same that was to be expected had it been the Power of Gravity alone, by which the Moon was there influenced'.[21] Or, as Pemberton put it, Newton's 'computation did not answer expectation'.[22] Both Whiston and Pemberton state that Newton's test was a failure, that the predicted result of the inverse-square law was only approximately or roughly the observed value, but we have seen that Newton in his later memorandum said that the theoretical and observed values agreed 'pretty nearly'. This expression "pretty nearly", like "approximately", is a valid translation of the Latin adverb "quamproxime", which Newton uses frequently in the *Principia*. The Newtonian phrase would make it seem as if the result had been "pretty good", and could thus have provided support for Newton's claim to have found and to have made a satisfactory test of the inverse-square law at least a decade before his correspondence with Hooke (in 1679/80). But the expressions used by Whiston and Pemberton convey the impression that Newton was dissatisfied by the result, which in fact did *not* confirm the theory.[23] Furthermore, both Whiston and Pemberton assigned a cause to the failure of Newton's "moon test"; it was chiefly that he had used a faulty value for the radius of the earth. We have seen that Newton did, in fact, use a value that was not the true one.[24]

In relating how Newton's "moon test" led to a poor result, Pemberton reported that Newton had then 'concluded that some other cause must at least join with the action of the power of gravity on the moon'. Whiston was more specific about this 'other cause', saying that his 'Disappointment . . . made Sir *Isaac* suspect that this Power was partly that of *Cartesius*'s Vortices'.[25] D. T. Whiteside's research has given independent confirmation to that part of Whiston's statement in which he said that Newton then believed in a lunar vortex.[26]

We conclude that the Newton of the 1660's was very far indeed from a concept of universal gravitation and a law of universal gravity. In the first place, his concept of a "conatus recedendi a centro", like his later "vis centrifuga", was not a mutual force between two bodies, but concentrated on a single orbiting body rather than on the mutual (equal and opposite) reaction of every pair of bodies. In order to get to universal gravitation, Newton had to get rid of both the Cartesian "conatus recedendi a centro" and the Huygenian "vis centrifuga" and to adopt a new concept of "vis centripeta".[27] He also had to abandon any ideas about balancing centrifugal and centripetal forces in a kind of equilibrium and to commit himself to a quite different system of analysis of orbital motion, now conceived as the continual action of a centripetal force deflecting a body from its straight-line inertial path so as to make it follow a curve. Furthermore, there was no hint as yet that there would be a force of attraction between every pair of bodies, wherever they might be. And although Newton was aware of the inverse-square relation (in the "conatus" or "endeavour"), he had not as yet made precise the concept of mass, which enters along with distance as a primary factor in the gravitation law. In short, none of the essential concepts necessary for the formulation of the idea of universal gravitation was present in Newton's thought in the 1660's, and there is no basis for discussing a "law" of "universal gravity" before the 1680's. The first turning point came in relation to his correspondence with Hooke in 1679–80, but it would not be until five years after that

that Newton would take the first intellectual leap forward toward the concept of universal gravity acting according to the law that made Newton world-famous.

4. PRELIMINARIES TO UNIVERSAL GRAVITY: 1679–1684

A major preliminary to universal gravity was Newton's abandonment of the idea of "conatus" or "endeavour" to move and the concomitant conceptual shift from centrifugal to centripetal force, a change that marked a transition from the earlier Cartesian philosophy to the new physics of dynamical central forces.[1] The new approach to orbital motion was based on the idea of a continual deflecting action of a centrally directed force on a body with an initial component of uniform rectilinear (i.e., inertial) motion. This approach is set forth in prop. 1 of Book 1 of the *Principia*, and has been the hall-mark of analytical celestial mechanics ever since Newton's day. Newton himself said more than once that it was the correspondence with Hooke during the years 1679 and 1680 that really set him thinking fruitfully in a new way about the cause of celestial motions.[2] In this exchange of letters, Hooke actually suggested to Newton that the proper way to analyze the orbital motion of planets should be in terms of a linear component of inertial motion and a continuous sun-directed force. This approach to orbital motion was stated unambiguously and explicitly by Hooke to Newton. At the time this correspondence began, Newton was still mired in very confusing older notions. Furthermore, it is evident that Newton had not been completely and at once won over to Hooke's viewpoint, since he continued to write about aethereal vortices even after this correspondence.[3]

I believe it is important to understand that Newton's debt to Hooke must have been on the level of conceptual analysis that I have just described – a point that has been made by Westfall, Whiteside, and other writers – because what Newton certainly did *not* get from Hooke in the first instance was the concept of a force varying as the inverse square of the distance. He was quite right in claiming that for this he would have had no need for Hooke.[4] But there can be no doubt that it was Hooke who insisted that celestial motions be solved by the continuous action of a sun-directed force, a force continuously deflecting planets away from their otherwise uniform rectilinear or inertial motions.

Hooke had written to Newton in 1679,[5] soon after he had become the Secretary of the Royal Society, suggesting a 'philosophical' correspondence, and asking Newton 'particularly if you will let me know your thoughts of a ... [hypothesis of mine] of compounding the celestiall motions of the planetts of a direct motion by the tangent & an attractive motion towards the centrall body'.[6] In reply, Newton said that 'I did not before the receipt of your last letter, so much as heare (that I remember) of your Hypothese of compounding the celestial motions of the Planets, of a direct motion by the tangent to the curve & of the laws and causes of springyness ...'[7] Newton then communicated to Hooke 'a fansy of my own about discovering the earth's diurnal motion', essentially the general problem of the trajectory of a body dropped from a height above a rotating earth.[8] Newton supposed that if the falling body could move "through" the earth, its "path" would be a spiral tending to the earth's center. Hooke's next letter pointed out that Newton had erred; Hooke stated his own conclusion that the path 'would resemble an Elleipse'.[9] Newton then sent Hooke an acknowledgement of the legitimacy of Hooke's correction. Such a body, he now said, 'will not descend in a spiral to the very center [of the earth] but circulate with an alternate ascent & descent made by its *vis centrifuga* & gravity alternately overballancing one another'.[10] This reply, among other things, shows that at that time Newton was thinking of planetary motions in terms of a balancing of centrifugal force and gravity. As Whiteside has observed, Newton was not as yet 'prepared either to accept that an ellipse might be traversed under some suitable law of gravitational decrease with distance ... or even to believe that the problem was of scientific importance'.[11]

Hooke would not give up his pressure on Newton, and in reply to Newton stressed the significance of the problem ('of greate Concerne to Mankind, because the Invention of the Longitude by the Heavens is a necessary Consequence of it'), and bluntly stated that

> my supposition is that the Attraction always is in a duplicate proportion to the Distance from the Center Reciprocall [*that is*, the attraction is always inversely as the square of the distance from the center], and Consequently that the Velocity will be in a subduplicate proportion to the Attraction [*that is*, the velocity will be as the square root of the attraction], and Consequently as Kepler supposes Reciprocall to the Distance.[12]

A few days later, Hooke returned to this question again:

> It now remaines to know the proprietys of a curve Line (not circular nor concentricall) made by a centrall attractive power which makes the velocitys of Descent from the tangent Line or equall straight motion at all Distances in a Duplicate proportion to the Distances reciprocally taken [*that is*, as the square of the distances inversely].[13]

Hooke then said to Newton:

I doubt not but that by your excellent method you will easily find out what that Curve must be, and its proprietys, and suggest a physicall Reason of this proportion . . .[14]

Newton's last letter to Hooke in this series did not refer to the challenge put to him by Hooke. He did not ever send a word to Hooke about his own analysis of the path that a body would follow if it were constantly deflected from an inertial linear path by a force directed to a center, and he did not respond in writing to Hooke's direct statement concerning elliptical orbits, or – more truly – orbits that would (in Hooke's words) 'resemble an Elleipse'. There was no apparent reaction, at least no public reaction, on the part of Newton, to Hooke's challenge about an inverse-square force producing such elliptical orbits. Our chief source that it was Hooke's stimulus that goaded Newton's creative genius into solving the problem of the forces producing elliptical orbits is a series of documents in which Newton later recounted the historic steps leading to his mastery of celestial dynamics.

In these autobiographical memoranda,[15] Newton left no doubt as to the importance of his correspondence with Hooke in directing him to the solution of the dynamics of elliptical orbital motion.[16] For example:

In the end of the year 1679 in answer to a Letter from Dr Hook then Secretary of the R. S. I wrote that whereas it had been objected against the diurnal motion of the earth that it would cause bodies to fall to the west, the contrary was true . . . . Dr Hook replied . . . that they would not fall down to the center of the earth but rise up again & describe an Oval as the Planets do in their orbs. Whereupon I computed what would be the Orb described by the Planets. For I had found before by the sesquialterate [3/2] proportion of the tempora periodica of the Planets with respect to their distances from the Sun, that the forces which kept them in their Orbs about the Sun were as the squares of their mean distances from the Sun reciprocally: & I found now that whatsoever was the law of the forces which kept the Planets in their Orbs, the areas described by a Radius drawn from them to the Sun would be proportional to the times in which they were described. And by the help of these two Propositions I found that their Orbs would be such Ellipses as Kepler had described.[17]

Here Newton says that following his correspondence with Hooke, he found that for any kind of central force (that is, a centrally directed or centripetal force: a force directed toward a point as center), a radius vector drawn to a planet from that point would always sweep out equal areas in equal times. Or, he discovered at this time ('now') the dynamical significance of Kepler's law of areas in general and combined this result with what he 'had found before', that Kepler's third or harmonic law implies a force varying inversely as the square of the distance. These two results enabled him to find that elliptical orbits would result from this inverse-square law.[18]

We have no dated documentary evidence as to the level of Newton's comprehension of the dynamics of celestial motions in 1680, at the termination of the correspondence with Hooke. We do know, however, how he analyzed such orbital motions a few years later, when he wrote up his first essay on celestial mechanics following Halley's visit to Cambridge in August 1684.[19] I see no reason to suppose that his approach to the subject in that essay or tract, later developed more fully in Book 1 of the *Principia*, does not essentially represent the state of Newton's thinking about planetary dynamics following the stimulus of Hooke's correspondence. We may, I believe, be reasonably confident that Newton developed his ideas in or about 1680 essentially along a line marked in the *Principia* by props. 1, 4, 6, and 11 of Book 1 (corresponding to theorems 1, 2, and 3 and problem 3 of the tract *De Motu*, presumably composed in November 1684 and revised almost at once).[20]

It is only fair, however, to add that while almost all Newtonian scholars would agree with the foregoing paragraph, our colleague R. S. Westfall argues that Newton's thoughts on planetary dynamics after the correspondence with Hooke rather follow an essay, 'On Motion in Ellipses'.[21] There is no external or documentary evidence that enables scholars to set the date of the composition of this essay earlier than the *Principia*, or – at least – earlier than *De Motu*,[22] and there is physical evidence that the actual paper on which a major version is transcribed was in circulation in the early or mid 1690's. Until or unless some new evidence may be found that will require an earlier date for 'On Motion in Ellipses', there is no compelling reason to assume it was not what it purported to be, namely, a simplified or popular approach to the Newtonian result concerning elliptical orbits written by Newton for John Locke,[23] and not a primitive approach to the subject, dating from a time before the composition of *De Motu*.[24]

Props. 1, 4, and 6, leading to prop. 11 of Book 1 of the *Principia*, constitute an analysis of elliptical orbital motions in the following sequence (corresponding to theorems 1, 2, and 3, leading to problem 3 of *De Motu*). Newton begins with the law of areas in relation to the law of inertia (*Principia*, Bk 1, prop. 1; *De Motu*, theor. 1). First Newton shows that if a body moves with pure inertial motion (uniform and rectilinear), a line drawn from the body to any point not on the path of motion will sweep out equal areas in any equal times. Thus was revealed the first stage of the as-yet-unsuspected connection between the Cartesian law of inertia and the Keplerian law of areas. Next, Newton again considers a body in inertial motion; but now, at regular intervals, an impulsive or instantaneous force or blow strikes the body, all these forces being directed toward a given point. The result is a polygonal path. The triangles whose bases are the sides of the polygon and whose common vertex is the point toward which the forces are directed will all have the same area; they are swept out in equal times. Let the time between successive blows become smaller and smaller. In the limit, as the distance or time between successive impulsive blows becomes indefinitely small, the polygon approaches a continuous curve, and the motion remains area-conserving.[25] In this way Newton demonstrates that the area law in orbital motion is a consequence of the combination of an initial linear uniform or inertial motion and a central force field. This result is, in itself, remarkable, because it showed the significance of the general area law in relation to the law of inertia and a central force field, independent of any conditions of force other than the central direction. It is worthy of note that for the first time, according to documented evidence, we may see Newton fully aware of the second law of Kepler, or the area law, as a relationship of *astronomical* significance.

In this connection, we must remember that in those days the area law was not universally used or even mentioned and discussed by astronomers. In its place there were various substitutes, among them one devised by Kepler himself, in which the orbital motion or position of a planet in its elliptical orbit could be determined by a radius vector rotating uniformly about the empty focus of the ellipse. Since this introduced a slight error (less than 8 minutes of an arc in celestial longitude per revolution in the case of Mars, but a more significant discrepancy for Mercury), a geometric correction factor was added by imbedding the ellipse in a circle of reference for which the axis major of the ellipse was a diameter of the circle. The planet's position in the ellipse was then determined as before, with a correction factor. A perpendicular is erected on the major axis of the ellipse (or line of apsides) passing through the intersection of the rotating radius vector and ellipse and continuing until it reaches the circle. A second radius vector is now drawn from the empty focus to the intersection of the perpendicular and the circle. The intersection of this second radius vector and the ellipse gives a corrected planetary position.[26] Newton was familiar with this type of approximation and even made calculations of planetary positions using such methods. But there is a real question as to whether Newton actually or consciously knew of Kepler's law of areas as a significant law of planetary orbital motion earlier than 1679/80.

This question must be phrased carefully in order to get a meaningful answer. First of all, there is no documentary evidence of any kind that Newton was consciously aware of this law as a principle to be used in solving planetary motions at any time before 1679.[27] On the other hand, he certainly had encountered the second law of Kepler. Of this there can be no doubt. For instance, this law appears in Mercator's treatise on astronomy of 1676, a book in which a summary of one of Newton's own letters – on the libration of the moon – is published.[28] But it is to be remembered that in this book Mercator did not use Kepler's area law in finding planetary positions, but used the methods I have just described, with a variant of his own.[29] Newton was also aware of an essay by Christopher Wren on the solution of "Kepler's problem" by means of a cycloid; this was published as a supplement to Wallis's tract on the cycloid. Although Wren knew that "Kepler's problem" had arisen from astronomical considerations, and in particular from attempts to apply the area law, a reader of Wren's essay would not necessarily become aware that the area law might be of significance for planetary astronomy,[30] since Wren composed the essay to read almost entirely like an inquiry into pure geometry. In all of the autobiographical memoranda that have been preserved, Newton refers to the proposition about the physical meaning of the law of areas as a discovery made at the time of his correspondence with Hooke (i.e., in 1679 or 1680), although on occasion he either carelessly or erroneously (or perhaps mischievously) puts the date at 1677 or 1676–1677.

As Newton says, it was the law of areas, combined with the analysis of circular motion, which gave him the key to finding the force that acts on bodies moving in elliptical orbits. At least, that is what he said in the autobiographical document quoted above. This is, in fact, what follows upon the analysis of the area law both in the tract *De Motu* and in the *Principia*, and – as has been mentioned earlier – may be presumed to be Newton's approach to the problem after his correspondence with Hooke.[31] In theorem 2 of *De Motu* (prop. 4 of the *Principia*[32]), Newton starts out with the fact that uniform circular motion is area-conserving with respect to the center and deduces (from the previous theorem or proposition) that there must be a force directed toward the center. Newton called this a centripetal force or a *vis centripeta*, later saying that he had given it this name, having in mind to honor Christiaan Huygens, who had called the contrary force *vis centrifuga*;[33] (Huygens was the only living scientist, to my knowledge, to whom Newton applied the adjective "summus").[34] Since the law of this force (D/T^2 or v^2/r) was known, the application of Kepler's third law showed that in orbits that could be considered circular, the centripetal force must vary as the inverse square of the distance.[35] This is theorem 2 of *De Motu*, prop. 4 of the *Principia*; it embodies a result that Newton had found some years before, now transformed so that it applied specifically to a force, rather than an "endeavour", and to a centripetal rather than a centrifugal force. Then, in theor. 3 of *De Motu* (equivalent to prop. 6 of the *Principia*[36]), Newton introduced another breakthrough, a new mathematical concept of the measure of a force.[37]

Armed with this set of three theorems or propositions, Newton is able to show that elliptical planetary motion (or any elliptical motion according to the law of areas) requires an inverse-square force.[38] In the tract *De Motu*, this appears as prob. 3; in the *Principia* it is prop. 11. In *De Motu*, prob. 3 is preceded by problems 1 and 2, dealing respectively with motion in a circle (to find the force tending to a given point in the circumference) and with motion in an ellipse (to find the force tending to the center of the ellipse).[39] We have no way of telling whether these two problems were solved prior to 1684, when the tract *De Motu* was composed. Possibly, in the first go, Newton went directly to what became problem 3 in *De Motu* (prop. 11, *Principia*), that elliptical orbits imply an inverse-square centripetal force.[40] This is followed (*De Motu*) by theorem 4: if the force varies as the inverse square of the distance, the square of the periodic times will be as the cubes of the transverse axes of the orbits. Again, we have no way of knowing for certain whether or not this general theorem would have been part of Newton's first solution of the problem of planetary orbital motions.[41]

In the history of the exact sciences, this sequence of four theorems represents a break in continuity or intellectual leap forward of the highest magnitude: the creation of a wholly new approach to celestial physics, based on the concept of centripetal forces. In a sense the Newtonian celestial dynamics was a realization of Kepler's ideal of basing celestial physics on causes, as it is expressed in the full title of Kepler's *Astronomia Nova* (1604).[42] But Newton's success was quantum jumps away from Kepler's intuitive ideas and was based on a revolutionary system of dynamics forged on new concepts of force and momentum and of mass and inertia; Newton was producing a new physics of which Kepler had had but at most a mere visionary glimpse. Neither Galileo nor Descartes had ever envisioned such a celestial dynamics. The Newtonian formulation left even Huygens far behind. There was nothing quite like this Newtonian physics in existence at that time in any writings that we know of – in print or in manuscript. And so we may easily understand why Newton was so annoyed when Hooke demanded credit for the law of the inverse square, as if his mere insight into the inverse-square relation had entitled him to a major share in the creation of modern physics.

In point of fact, there is an oddity in the letter in which Hooke advanced his suggestion to Newton that the law of planetary force varies as the inverse square of the distance. For Hooke drew attention to a curious consequence of his supposition 'that the Attraction always is in a duplicate proportion to the Distance from the Center Reciprocall': he said that, as a result, 'the Velocity will be in a subduplicate proportion to the Attraction, and consequently as Kepler supposes Reciprocall to the Distance'.[43] In other words, $F \propto 1/r^2$ and $v \propto \sqrt{F}$; hence $v \propto \sqrt{1/r^2}$ or $v \propto 1/r$. That is, Hooke based his claim for the inverse-square law on a letter which assumed that another law of Kepler's, other than the area law, follows from the inverse-square law, namely, that the speed is inversely proportional to the distance of the planet from the sun. As our colleague E. J. Aiton has shown,[44] Kepler himself had at one time propounded this speed law but had later rejected it in favor of the area law, with which it is of course inconsistent. In the *Principia*, Newton showed in prop. 16 of Book 1 what the true law of planetary speed is: that the speed is not inversely proportional to the direct distance from the sun to a planet, but rather to the perpendicular distance from the sun to the tangent to the orbit at the planet's position.[45]

By the time of the second edition of the *Principia*, Newton had found that he could introduce this important result somewhat earlier in the text than prop. 16, in section 2 of Book 1. He then shifted the original corollaries to prop. 1 so that they would become corollaries to prop. 2. This made it possible for him to introduce a set of brand new corollaries to prop. 1. From then on, the first corollary to the first proposition of the first Book of the *Principia* showed that Hooke's supposition was false, that the true law of speed is the Newtonian one.[46]

5. THE ESSENTIAL STEP TO UNIVERSAL GRAVITY IN 1685

Newton's treatment of Kepler's law makes it possible to illustrate an aspect of the *Principia* and its central concept of universal gravity that is not generally understood: to set an accurate date for the earliest possible time when Newton could have begun to think about universal gravitation, the time when Newton took the first step that made the concept of universal gravitation possible (and probably, considering Newton, made that result inevitable). That date is soon after November 1684. We know that this is so because by November 1684, he had worked up the four propositions forming the main sequence that I have discussed in §4, as a part of a draft of the tract *De Motu*, which he sent on to London for Edmund Halley.[1] In this early draft, Newton (prob. 3, corresponding to prop. 11 of the *Principia*) proudly exhibits the derivation of what he there calls the law of centripetal force directed to a focus of an ellipse on which an orbiting body moves (according to the law of areas, since it is moving under the action of a central force). This force varies 'inversely as the square of the distance'. In *De Motu*, this leads Newton to the following conclusion:

> Scholium: Therefore the major planets revolve in ellipses having a focus in the centre of the sun, and radii drawn [from planets] to the sun describe areas proportional to the times, entirely [or exactly] as Kepler supposed.[2]

Had Newton actually proved this scholium? Certainly not. Did he long continue to believe it? Equally not. In fact, it is strictly false.[3] It is false because it does not take into account that the planets must attract the sun if the sun attracts them, from which it would logically follow that the planets must also attract one another. In this case, as Newton would shortly realize, there would not be simple Keplerian orbits in ellipses about the sun as focus according to the law of areas. Since interplanetary attractions are a special case of universal gravitation, and since Newton was not aware of such interplanetary attractions in November 1684, when he sent a copy of his tract *De Motu* to the Royal Society, it follows that he had not as yet conceived of universal gravitation. We shall see that a universally acting gravitating force could not have been conceived before certain revisions of *De Motu*, which were made either in December 1684 or in the very beginning of 1685,[4] almost certainly the latter.

A close reading of the *Principia*, or of the tract *De Motu*, shows that Newton could not have demonstrated that the planets move 'entirely' or 'exactly' (*omnino*) according to these two laws of Kepler. Newton proved something very different, namely, that Kepler's area law and law of ellipses are true only for what is essentially a one-body system,[5] that is, a single mass point moving with an initial component of inertial motion in a central gravitational field. Then Newton recognized that this one-body system does not correspond to the world of experience, but represents nature simplified; it is a mental construct useful for the mathematical investigation of artificial systems, but it is not nature herself. The key to the next stage in Newton's thought was to see the consequences of what we know today as Newton's third law of motion, perhaps the most original of his three laws,[6] the generalization of action and reaction. Testimony to the degree of originality of this law is given by the fact that it has been, and still is, so often misunderstood in its applications, especially by those who conceive it in relation to a condition of equilibrium.[7]

The one-body system reduced the earth to a mass point and the sun to a mere center of force.[8] But, as Newton explained the matter in the *Principia*, he had – as a consequence of having adopted this one-body system – limited his considerations to the artificial conditions of 'motions of bodies attracted toward an unmoving center' and this 'hardly exists' in the real world.[9] For the fact of the matter is, he said, that 'attractions customarily are directed toward bodies and – by the third law of motion – the actions of attracting and attracted bodies are always mutual and equal'. Hence, he continued, 'if there are two bodies, neither the attracting nor the attracted body can be at rest'. Rather, 'both bodies (by the fourth corollary of the laws) revolve about a common center, as if by a mutual attraction'. These statements, from the introduction to sect. 11 of Book 1 of the *Principia*, set forth unequivocally the line of development of Newton's thinking after completing the first draft of *De Motu*, that is, no earlier than December 1684 and most likely early in 1685.[10]

In other words, Newton has seen that a consequence of the third law of motion is that if the sun pulls on the earth, then the earth must also pull on the sun and with a force of equal magnitude. The result is, therefore, that in such a two-body system, the earth will not move in a simple orbit around the sun; rather the sun and earth will each move about their mutual center of gravity. In the transition from the mental construct of a one-body system to the realities of the experientially perceived external world, the original conclusion from prob. 3 of *De Motu* [prop. 11 of the *Principia*] proves to be inadequate. But there is a further consequence of the application of the third law of motion. The considerations of the sun and each planet as a system of two mutually attracting and attracted bodies lead to the result that every planet is not only an attracted body (attracted by the sun) but is also a center or origin of attractive force (attracting the sun). It follows that each planet should not only attract and be attracted by the sun, but should also attract and be attracted by all the other planets. That is, there is a logical step from an interactive two-body system to an interactive many-body system.[11]

We may see these steps as they actually occur in a revision of the tract *De Motu*, apparently made early in 1685 (or, just possibly, in the last days of 1684). Newton now concludes that because of 'the actions of all of them on one another' ('eorum omnium actiones in se invicem'), 'the planets neither move exactly in ellipses nor revolve twice in the same orbit'. And this leads him to the following result:

So that there are as many orbits to a planet as it has revolutions, as in the motion of the Moon, and the orbit of any one planet depends on the combined motion of all the planets, not to mention the action of all these on each other.[12]

He then says that

to consider simultaneously all these causes of motion and to define these motions by exact laws allowing of convenient calculation exceeds, unless I am mistaken, the force of the entire human intellect.

He confesses that only by 'ignoring those minutiae' will 'the simple orbit and the mean among all errors ... be the ellipse of which I have already treated'.[13]

In the *Principia*, begun in 1685 as an expansion of the tract *De Motu*, we can see similar considerations being applied to the satellites of Jupiter and to our moon (props. 4, 5, Book 3). The third law of motion, applied to these two cases, leads to the conclusion that there is a mutual force between the earth and moon and between Jupiter and each of Jupiter's satellites. Hence, by the same argument used for planets, it is concluded that our moon (and each of Jupiter's satellites) is both the origin of a force of attraction and an object on which such attraction acts. It follows that the satellites attract one another, that they and the moon are attracted by (and equally attract) the sun and each of the planets. Eventually, the same argument will lead to the result that an apple and any other terrestrial object must attract the earth and that all terrestrial objects will attract one another. And it is but a short step from that point to the concept of universal gravitation. Here prop. 6 states that 'all bodies gravitate toward every planet' and that the 'gravities of [any] bodies toward any single planet, at equal distances from the planetary center, are proportional to the quantity of matter in the individual bodies'.

In this progression, there are a number of steps that transcend the simple logic of this presentation. First of all, there is the problem of whether or not the force of attraction in the earth and moon, all the planets, their satellites, and the sun is one and the same, a set of different manifestations of terrestrial gravity. Second, there is an even more basic problem: can there exist such "grasping" forces which can extend through "empty" space over millions of miles to affect a distant body such as a planet or a comet? As to the first of these, in the *Principia* Newton shows by the "moon test" (prop. 4 & schol., bk. 3) that the force of the earth's gravity extends to the moon. That is, he proves that if the moon's orbit results from the moon's constant falling away from a linear inertial (or tangential) path, then the rate of fall is exactly the same as the rate of fall of a body moved from the earth's surface out to the moon's orbit, assuming that the earth's gravity would still pull that body toward the earth and that this force varies inversely as the square of its distance from the earth's center. Since the moon is 60 earth-radii away from the earth's center, the force of terrestrial gravity, or the rate of free fall that force produces, should be 1/3600th of what it is at the earth's surface or 1 earth-radius away from the earth's center. And this is just the rate at which the moon "falls".

Next Newton considers the satellite system of Jupiter (prop. 5). Then (prop. 5, coroll. 1, 3; prop. 6–7) he leaps to the conclusion that the earth-moon force, the Jupiter-satellite force, the sun-planet force, and hence the forces among planets and satellites, are the same as the cause of terrestrial gravity and that hence they may all be called "gravity". The reason is that they are all bodies of the 'same kind', that is, they must all have mass (which is the determining factor of the bodies in sun-planet attraction or in planet-satellite attraction). As to the more daring position that there are such forces of attraction, there are no simple documented answers, but – as shall be explained in §6 immediately following – the "Newtonian style" made it possible for Newton to work out the mathematical principles of celestial dynamics without being unnecessarily impeded by the fact that the received "mechanical" philosophy did not have any place for such forces, and was restricted to contact forces such as pressure or the forces of impact arising from matter and motion.

There are no documents that indicate exactly how, in the month or so between writing the first draft of *De Motu* and revising it, Newton came to perceive that the planets act gravitationally on one another. Nevertheless, the passage cited above expresses this perception in unambiguous language: *eorum omnium actiones in se invicem* ('the actions of all of them on one another'). A direct consequence of this mutual gravitational attraction is that all three of Kepler's laws are not strictly true in the world of physics but are true only for a mathematical construct in which point masses that do not interact with one another revolve in orbit about either a mathematical center of force or a stationary attracting body. The distinction Newton draws between the realm of mathematics, in which Kepler's laws are truly laws, and the realm of physics, in which they are only 'hypotheses,' or approximations, is one of the revolutionary features of Newtonian celestial dynamics.[14]

I have assumed that the third law of motion was the key factor in the reasoning that led Newton to suggest mutual gravitational perturbations of planetary orbits. There is no direct evidence for my assumption because no documents exist in which there is an antecedent version of his statement: the actions of all them on one another. Nevertheless, there is strong indirect evidence. In the spring of 1685, a few months after revising *De Motu*, Newton was well on his way to finishing the first draft of the *Principia*. In the initial version of what was to become the third book of the *Principia*, 'The System of the World', he spelled out the steps that led him to the concept of planetary gravitational interactions. In these steps the third law of motion has the chief rôle, and I see no reason to believe these are not the very steps that led him to the same concept a few months earlier, when he was revising *De Motu*.

Here are two passages from the first draft of 'The System of the World' (as translated from the Latin by Anne Whitman and I. B. Cohen in our forthcoming version) that bring out the crucial rôle of the third law of motion:

20. *The agreement between the analogies*

And since the action of centripetal force upon the attracted body, at equal distances, is proportional to the matter in this body, it is reasonable, too, that it is also proportional to the matter in the attracting body. For the action is mutual and causes the bodies by a mutual endeavor (by law 3) to approach each other, and accordingly it ought to be similar to itself in both bodies. One body can be considered as attracting and the other as attracted, but this distinction is more mathematical than natural. The attraction is really that of either of the two bodies toward the other, and thus is of the same kind in each of the bodies.

21. *And their coincidence*

And hence it is that the attractive force is found in both bodies. The sun attracts Jupiter and the other planets, Jupiter attracts its satellites, and similarly the satellites act on one another and on Jupiter, and all the planets on one another. And although the actions of each of a pair of planets on the other can be distinguished from each other and can be considered as two actions by which each attracts the other, yet inasmuch as they are between the same two bodies they are not two but a single operation between two termini. Two bodies can be drawn to each other by the contraction of one rope between them. The cause of the action is twofold, namely, the disposition of each of the two bodies; the action is likewise twofold, insofar as it is upon two bodies; but insofar as it is between two bodies it is single and one. There is not, for example, one operation by which the sun attracts Jupiter and another operation by which Jupiter attracts the sun, but one operation by which the sun and Jupiter endeavor to approach each other. By the action by which the sun attracts Jupiter, Jupiter and the sun endeavor to approach each other (by law 3), and by the action by which Jupiter attracts the sun, Jupiter and the sun also endeavor to approach each other. Moreover, the sun is not attracted by a twofold action toward Jupiter, nor is Jupiter attracted by a twofold action toward the sun, but there is one action between them by which both approach each other.

Next Newton concluded that 'according to this law all bodies must attract each other'. He proudly presented the conclusion and explained why the magnitude of the attractive force is so small that it is unobservable in ordinary terrestrial bodies. 'It is possible', he wrote, 'to observe these forces only in the huge bodies of the planets'.

In Book Three of the *Principia*, which is also concerned with the system of the world but is somewhat more mathematical, Newton treats the topic of gravitation in essentially the same way. First he extends the concept of a weight force, or terrestrial gravity, to the moon and proves by the moon-test that the force varies inversely with the square of the distance. Then he identifies this same terrestrial force with the force of the sun on the planets and the force of a planet on its satellites. All these forces he now calls gravity. With the aid of the third law of motion he transforms the concept of a solar force on the planets into the concept of a mutual force between the sun and the planets and between one planet and another. Similarly, he transforms the concept of a planetary force on the satellites into the concept of a mutual force between planets and their satellites and between satellites. The final transformation is the notion that all bodies interact gravitationally.[15]

It was a momentous step from Newton's concept of an interactive two-body system to an interactive many-body system. The logic of this line of reasoning is universal. All matter of, on, or near the earth is characterized by having mass (and other properties). Such matter has weight, that is, is drawn, pulled, or impelled toward the earth. But, by the same third-law argument used for sun and planets, and for planets and satellites, it follows that all such bodies (or samples of matter) are both centers of attractive force and objects on which the attractive force acts; hence there must be a gravitational action between any two bodies. In this way, Newton's daring and genius generalized the inverse-square law of force for the actions of the sun on each planet and extended it to all matter so as to become the law of universal gravitation.

My analysis of the stages of Newton's thinking should not be taken as diminishing the extraordinary force of his creative genius; rather, it should make that genius plausible. The analysis shows Newton's fecund way of thinking about physics, in which mathematics is applied to the external world as it is revealed by experiment and critical observation. This way of thinking, which I call the Newtonian style, is epitomized in the English title of Newton's great work: *Mathematical Principles of Natural Philosophy*.

6. THE "NEWTONIAN STYLE"

The *Principia* discloses a style of mathematical physics that may indeed prove to be the most revolutionary feature of that great treatise.[1] The Newtonian style has also been a cause of much misunderstanding concerning Newton's intentions. Newton himself never set forth the principles of this style of science in a series of formal precepts; its essential features have to be gathered from an analysis of his procedure. In introducing the expression "Newtonian style", I am not implying anything more than a description of the way in which Newton combined two levels of ontology, the mathematical and the experiential. A case could perhaps be made that this style is Galilean or Keplerian, rather than an invention of Newton's. In fact, Edmund Husserl has written at large concerning the "Galilean" style,[2] essentially the mode of modern mathematical physics; from this point of view, the Newtonian style can be seen as a highly advanced and very much refined development of the Galilean.[3]

The features of the Newtonian style can be seen easily in his treatment of Kepler's laws. In *De Motu* and in the *Principia*, Newton begins by exploring the conditions under which Kepler's laws are true. We have seen how Newton demonstrates that a pure elliptical orbit results from what may be called a one-body system: essentially a mass point and an inverse-square force constantly directed to a central point.[4] The body in question in the opening propositions of *De Motu* and the *Principia* is a mass point; only later on in the *Principia* does Newton address himself to real bodies in the physical sense of their sizes, shapes, and compositions. The results hold equally, of course, for an ensemble of such one-body systems, all having a common force center, so long as the bodies are not interactive. It is obvious, then, from Newton's analysis that the Keplerian first two laws of planetary motion are exact or true only for a highly artificial system: a mental construct that represents a mathematicization of nature simplified. Furthermore, Newton's treatment of Kepler's laws occurs in a mathematical space, in which what may at first glance appear to be physical lines turn out to be infinitesimal line-elements and even – as D. T. Whiteside has shown – infinitesimals of higher order than one.[5]

In the *Principia*, Newton explores the consequences of this mental construct by means of mathematics – geometry, proportions and what is essentially algebra, the process of limits, fluxions, and so on. Newton is fully aware that what he has been exploring in this manner is not nature but mathematics, the mathematics of the system he has imagined or contrived, in which Kepler's laws are posited as true. Hence, because he is ultimately interested in physical nature, in the external world whose properties are revealed to us by experiments and observations, he must compare the mental construct and the laws or rules that hold in it with the astronomical universe. But in the early part of the *Principia*, as Newton declares in the introduction to sect. 11, Bk 1, he has been studying 'the motions of bodies attracted toward an unmoving center' although such a condition 'hardly exists in the real world'. In order to make a comparison of the mathematical system he has developed and the physical universe, he must, accordingly, translate the elements of the mental construct into their physical equivalents. Thus the geometric space, mathematical time, and mass points become translated into physical space, phenomenological time, and astronomical bodies having mass. This is the second phase in the Newtonian style, the transfer of the results he has obtained in mathematics to physical nature. The first phase is to develop the mathematical consequences of the mental construct, derived in the first instance from a simplification of natural conditions, followed by a mathematicization. In phase one, it is also possible to explore the mathematical consequences of conditions which are arbitrary and perhaps only of mathematical interest. Thus Newton can be interested in Kepler's third law ($T \propto r^{3/2}$) which, for circular orbits, yields a force $F \propto 1/r^2$, as in corol. 6 to prop. 4 of Bk 1, but he is equally free to explore any condition of T being proportional to r (that is, T may be proportional to any power r^n of the radius), whereupon the velocity becomes proportional to $1/r^{n-1}$ and $f \propto 1/r^{2n-1}$.

The comparison of the mental construct of a one-body system and the world of physical nature leads Newton from consideration of a mass point and a central force to a two-body system, in which two bodies or two mass points mutually attract each other. This introduces a modification of the original mental construct; phase two has produced a new or revised phase one, in which Newton once again applies his mathematical techniques to derive a set of results on a higher level of complexity than the original ones. Again, these are compared with phenomena, or with the rules or laws derived from phenomena, in another phase two. It is now seen that even the two-body system is too limited. Hence Newton introduces a third, or perturbing, body, and goes on to explore the mathematical consequences of adding this condition to the mental construct.[6] In a successive alternation between a phase one and a phase two, Newton will introduce many bodies,[7] bodies with definite and significant sizes and shapes,[8] and mediums through which bodies may move.[9] He will also consider motion under varieties of constraints, such as along particular surfaces.[10] In this fashion, Newton actually develops mathematical principles in a mathematical context, many of which are analogues of physics and have the same names as in physics (motion, path, body, etc.). For Newton there is a final phase three, in which these principles will no longer be purely mathematical but will be applied to the real world of physical nature as revealed by experiment and observation. The purely mathematical principles then become the 'mathematical principles of natural philosophy', which is the full title of the *Principia*.

Books 1 & 2 of the *Principia* are largely concerned with what I have called here phases one and two. Book 3 illustrates phase three, by using the 'mathematical principles of natural philosophy' in order to develop the 'system of the world'.[11] In the introduction to Book 3, Newton begins by stating explicitly that the 'principles' that he has presented in 'the preceding books . . . are . . . strictly mathematical'; yet they are 'those on which natural philosophy can be based'. Of course, even though his constant aim has been to elaborate 'mathematical principles' of use for 'natural philosophy' – i.e., to discover 'laws and conditions of motions and forces, which especially relate to [natural] philosophy' – he has not absolutely restricted himself in this way. As a creative mathematician he could hardly have done so. Yet, from time to time, in order to remind his readers that the 'principles' were not 'barren', he had introduced 'some philosophical scholiums, treating topics that are general and seem to be the principal foundations of [natural] philosophy, such as the density and resistance of bodies, spaces void of bodies, and the motion of light and sounds'.[12]

For Newton there are thus two wholly distinct levels of discourse in the *Principia*: the mathematical and the physical. In prop. 48, Bk. 2, Newton refers to a proportion that is not as exact as it would be in a world of mathematics ('Accurata quidem non est haec proportio'). In the world of physics, such inexact proportions abound. In this particular case, he says that unless certain contractions and expansions of a given elastic fluid are not extremely great, the proportion 'will not be wrong so far as the senses can perceive, and so it can be considered as physically exact' ('. . . non errabit sensibiliter, ideoque pro physice accurata haberi potest'), even though it is not mathematically exact. This is, I believe, the sense in which Newton considered Kepler's planetary laws: they may (in the above sense) 'be considered as physically exact' relations in the physical solar system, but they are not truly exact. They are truly exact or absolutely true only in an artificial world created in the mind, such as the mathematical system of one body and a force directed toward a center. In the second and third editions of the *Principia*, at the beginning of Book 3 ('On the System of the World'), Newton introduced

Kepler's law of areas and harmonic law for the physical world of the 'primary planets' and their satellite systems under the rubric of 'PHAENOMENA', by which he evidently meant that these are rules that are 'physically exact' to within a certain degree of observation. And he gave numerical data to indicate the limits of exactness of these phenomenological rules. In the first edition, these rules were not called 'PHAENOMENA' (as in the later editions) but were 'HYPOTHESES'.[13] This word "HYPOTHESES" was evidently intended to convey strongly that these were only phenomenological statements or rules, which were not true or 'exact' in an absolute or mathematical sense. As 'hypotheses', Kepler's planetary rules could be used freely without any commitment to their absolute truth or exactness. In this sense there is an equivalence between 'hypotheses' and 'phaenomena'. In the first edition, however, toward the end of Book 2, Newton referred to the 'Hypothesis 4' of Book 3 (stating the harmonic law for Jupiter's satellites) as follows: 'For it is a phenomenon that the periodic times of the planets revolving about Jupiter are in the sesquialteral [3/2] ratio of their distances from the center of Jupiter.'[14] In his own mind, in 1686–1687, the planetary "hypotheses" of Book 3 were already conceived as "phenomena" in the equivalent sense of these terms just indicated.

The phase one in Newton's *Principia* was not restricted to the motions of bodies with respect to central forces.[15] Other examples are systems of mutually repelling particles that ultimately yield a model for Boyle's law,[16] particles whose motion simulates light,[17] wave motion in fluids,[18] motion along various types of surfaces,[19] different kinds of pendulum vibration,[20] general conditions of force that go far beyond the two major ones in nature (the inverse square of the distance and the direct distance), and a variety of different conditions of resistance to motions in fluids.[21] All of these occur in Books 1 & 2, 'On the Motion of Bodies', the books of the *Principia* which he says are mathematical – with no supposition being made as to the reality (or even the possibility of reality) of the forces invoked. This is in contrast to Book 3, 'On the System of the World', in which Newton shifts from mathematics and considerations of imagined mathematical systems or mental constructs to natural philosophy. In Book 3 he is no longer dealing with arbitrary mathematical conditions, or even with nature simplified, but with the world of directly observed phenomena. In Book 3, however, the forces do have two mathematical aspects. First, they follow mathematical "rules" (as being inversely proportional to the square of the distance). Second, universal gravity may be considered to be "mathematical" to the degree that Newton does not inquire into its nature, cause, or mode of transmission.

The Newtonian style thus encompassed two usually distinct and clearly discernible levels of discourse and meaning in the *Principia*. One was mathematical and the other physical: one invoked mathematical principles in relation to mathematical systems and mental constructs, the other applied these principles directly to natural philosophy. A critical reading of the *Principia* shows how extremely careful Newton was in trying to keep these two levels distinct, although many commentators from that day to ours have confused them.

On this aspect of the Newtonian style, there is no want of direct evidence. A striking example occurs in a letter to Halley, written on 20 June 1686, a couple of months after Book 1 of the *Principia* had been completed and sent off for publication. In this letter Newton mentions that in the 'small ascent & descent of projectiles above the earth, the variation of gravity is . . . inconsiderable'; in fact, it is so minuscule that it may be neglected. Hence, Newton says, in considering projectile motion, it is customary to use 'the vulgar Hypothesis' of 'uniform gravity'. And then the Newtonian style is epitomized in these words: 'And why might not I as a Mathematician use it [i.e., this vulgar hypothesis] frequently without thinking on the philosophy of the heavens or believing it to be philosophically true?'[22] Here, then, is an example of the Newtonian style in action, with a clear distinction being made between what is mathematical and what is – in Newton's own words – 'philosophically true'.

Again and again in Book 1 and Book 2, Newton says clearly that he is dealing with mathematics. For example, in sect. 7, Bk. 1, he concludes a discussion of centripetal forces – both those that vary directly as the distance and those that vary inversely as the square of the distance[23] – with a mathematical generalization concerning the motion of bodies on the supposition of 'a centripetal force of any kind'. In the succeeding sect. 8 (where Newton assumes the ability to perform integration – as he says, 'granting the quadratures of curvilinear orbits'), there occurs prop. 41, the crown of all: 'supposing a centripetal force of any kind, . . . to find the trajectories [i.e., the curves] in which bodies will move, and the time of their motions in the trajectories so found'. This is a mathematical result of great generality and power, independent of any specified physical conditions of force.[24]

That the first sections of Book 1 are either primarily mathematical or deal exclusively with mathematical systems or mental constructs is evident from the general lack of what Newton has designated as 'philosophical' or physical scholiums (there are none at all in sects. 7–8–9–10), although plainly Newton is developing propositions concerning the motion of bodies under the actions of forces or constraints that are to be of service somewhat later on in solving problems in the physics of motion. In other words, he is dealing with the mathematics of limited or arbitrary conditions; or, he is exploring the mathematical properties of artificial situations or mathematical systems.

In the introduction to section 11 of Book 1, to which I have referred several times, Newton says that he has thus far been concerned with 'the motions of bodies attracted toward an unmoving center, which, however, hardly exists in the realm of nature' ('quale tamen vix extat in rerum natura'). This is a clear and unambiguous statement that Newton has been thus far exploring a purely mathematical system, a mental construct rather than phenomenological nature. But now, he says, he will turn to bodies that attract each other mutually, according to the third law of action and reaction, as in the world of physics. Here we see phase two leading to a new phase one. From here on, the expression "centripetal force" will no longer be appropriate, since there is no longer a fixed center. In what follows, he will consider the former 'centripetal forces as attractions, although perhaps – if we were to speak in the language of physics – they might better be called impulses'. But here (phases one and two), he says, 'we are concerned with mathematics'; that is, we are not engaged in natural philosophy or physics, as will be the case later on in Book 3. Indeed, he says, he is 'putting aside any debates concerning physics', that is, he is not at all concerned with any physical aspects of the mutual forces which impel or attract or in any way draw one body toward another. Such considerations need not be of concern in phases one and two. When he uses the term 'attraction' to denote these mutual forces, it is merely for convenience, as he says, 'so as to be more easily understood by mathematical readers'. That is, his whole purpose here in phases one and two is mathematical, and he is using a familiar word that best and most simply expresses the concept he is developing mathematically. Had this not been phase one, he might have written of "mutually impelling forces [or impulses] that tend to draw two bodies together" or he might have used some other complex locution that would not have had such potentially objectionable physical overtones as the word "attraction". But to get directly at the mathematical relations, and in language as simple as possible, he had chosen the most expressive single word for the mathematical concept he was invoking, namely, "attraction".[25] From the point of view of the Newtonian style, Newton's use of the word "attraction" in the introduction to section 11 is unambiguous. But critics from Fontenelle to Alexandre Koyré[26] have assumed that Newton was making "attraction" in this mathematical context (phases one and two) equivalent to universal gravity in a physical context (as in phase three, Book 3). In Book 3, as a matter of fact, the word "attraction" scarcely appears. It is almost exclusively reserved for Books 1 & 2, save in the context of electrical and magnetic attractions.[27]

Newton actually explains these three phases of what I have called the Newtonian style in the conclusion to sect. 11, Book 1, of the *Principia*. Here he says, 'In mathematics, the task is to investigate the quantities of forces and those ratios that follow from any conditions that may be supposed'. I have called this phase one; it is tempered by a phase two, in which physical considerations cause revisions of the conditions of phase one, yielding a new phase one, which itself may be tempered by a new phase two, and so on. This phase two is also described by Newton in the conclusion to sect. 11 as follows: 'Then', he says, 'moving on to physics, these ratios must be compared with the phenomena, so that it may be found out what conditions or laws of forces apply to each kind of attracting bodies'. After this, phase three: 'then, finally, it will be possible to dispute more safely about the physical species, physical causes, and physical proportions of these forces'. Newton makes this same point very effectively in the discussion of the Definitions at the beginning of the *Principia.*[28]

The whole of the *Principia* (the opening sections of Definitions and Laws of Motion; Books 1, 2, & 3) exhibits the three phases of the Newtonian style that I have been describing. It is a feature of this style that only *after* Newton has exhibited the system of the world in phase three – retrodicting the known phenomena and predicting new and verifiable ones (for example, the perturbation of Saturn's orbit by Jupiter, and an aspect of the motion of the moon) – does he have to consider any questions of the cause, nature, and mode of transmission of universal gravity as a kind of sequel to phase three or as a supplement to his unfolding of the system of the world. The great power of the Newtonian style is that in phases one and two of developing the mathematical consequences of mental constructs, he did not have to introduce (at this stage) any questions as to the reality or the possibility of existence, or the degree of acceptability (according to the canons of the reigning philosophy) of the forces or other "physical" entities which he was considering "mathematically". He was not inhibited by such questions.[29] He could postpone them until a later time, that is, until he had proceeded to more and more complex systems, and had found at last (phase three) that the results could be applied directly to explain the phenomena of the external world of nature. These questions were obviously of great personal importance to Newton, as they were to his contemporaries. But the Newtonian style differentiated the science of the *Principia* from the received or common rules of science. The then-current canons of philosophy would not permit relegating all such questions to a supplementary stage. To many of Newton's contemporaries, the Newtonian system of the world could not be considered acceptable, since it was based upon a force of universal gravitation that could not be reduced to principles of matter and motion. The degree to which the Newtonian style was revolutionary may be seen in the fact that Newton's critics rejected it or could not understand it. They so strongly objected to the concept of a force acting at a distance that they would have nothing to do with what they saw as a system based on a return to the "occult qualities" which had supposedly been cast out of natural philosophy. Newton tried to answer his critics in the concluding General Scholium, introduced as a sequel to the *Principia* in the second edition (1713). He had produced 'mathematical principles' to serve for 'natural philosophy' (phases one & two) in Books 1 & 2, he said, without overt concern for ontological considerations of central forces. Now, in Book 3 (phase three), he maintained, he had established the existence in phenomena of effects directly attributable to the gravitational forces whose properties had been disclosed mathematically during phases one and two. Hence Newton could conclude, in the General Scholium, that 'gravity really exists' and that it 'acts according to the laws that we have set forth', and that it 'suffices to explain all the motions of the heavenly bodies and of our sea'. This 'is enough', ('satis est'), he asserts, to make gravity acceptable, even though he has 'not as yet assigned a cause to gravity'. What this implies is that an inquiry into the cause of gravity, or its mode of action or transmission, is supplementary to phase three and not at all an essential part of it. Newton thus proposes that his gravitational natural philosophy should be considered acceptable even though it does not go beyond phase three.

But even though Newton set forth the doctrine of a natural philosophy being acceptable without an understanding of the central concept, universal gravity, that did not mean that he himself was not vitally interested in what I have called the sequel to phase three, the search for cause. We know that Newton embraced Fatio's explanation of gravity as caused by the motion of aethereal particles;[30] he was greatly taken with the possibility that electrical phenomena might give him a key;[31] and he returned to an explanation based on an aether of varying density that would have an 'elastick force' which 'may suffice to impel Bodies from the denser parts of the Medium towards the rarer, with all that power which we call Gravity'.[32] By such a search for a cause of gravity, Newton hoped to find a reconciliation between the precepts of the "mathematical philosophy" and the concept of a grasping force that could stretch out over vast distances of apparently empty space. His aim was to reduce the actions of an "attractive" force of universal gravity to an effect caused by matter and motion, so that his system of the world would ultimately conform to the received philosophy of nature.

7. NEWTON'S VIEWS ON THE AETHER AT THE TIME OF WRITING THE *PRINCIPIA*: THE PENDULUM EXPERIMENT TO TEST THE RESISTANCE OF AETHER

Early in his career, Newton was interested in the observed fact that in a vacuum a freely swinging pendulum slows down and comes to rest; in fact, it does so almost as quickly in an evacuated vessel as it does in ordinary air. Newton interpreted this experiment as proving that there must be an aether: it is something that resists motion and that remains in a receiver from which the air has been pumped out.[1] In the *Principia*, Newton reported a different kind of experiment: to test whether a "dense" aether of the traditional kind, capable of producing "mechanical" effects on bodies, can exist. A long pendulum, hanging by a cord eleven feet long, had a bob which was hollow, so that it could be filled with different substances. Newton studied the way in which the pendulum, oscillating freely, would slow down when the bob was empty of anything but air and when the bob was filled with metal. Newton concluded from his experiments that in the latter case the resistance to motion arising from any 'aethereal and exceedingly subtle medium that quite freely permeates the pores and passages of bodies is either nil or completely insensible'.[2] In presenting this experiment in the *Principia*, Newton said that he was reporting the experiment 'from memory', since the 'paper on which I had once described it is lost'. He gives no hint as to whether the experiment had been done around 1679/80 or in 1684 or even while writing the *Principia* in 1685. I believe that the experiment was not performed until 1685.[3]

The subject of Newton's belief in an aether is of real importance in assessing the stages toward universal gravitation, because the concept of gravity being caused by an aether does not lead to *universal* gravitation, but even turns out to be in one respect inconsistent with universal gravitation. That is, the explanations of gravity by means of an aether could not account for a gravitational force acting mutually and equally on two gravitating bodies according to the inverse-square law. Newton's suggestion of a "shower" of aethereal particles[4] and his hypothesis of the mid-1670's, in which there would be a density gradient in the aether,[5] might supply a plausible mechanism for bodies being "heavy" toward the earth or toward the sun or any other planet, but could not explain why the earth would be just as "heavy" toward an apple as an apple is toward the earth. Furthermore, the supposition that the aether whirls in huge vortices, in the Cartesian manner, may serve to explain a gravitational "endeavour" of bodies in the vortex toward the center, but such an "endeavour" would not then depend on the presence of an actual material body physically located at the center, which could be the seat of that gravitational "endeavour". This is the reason – namely, that the vortex theory did not lead to the concept of a *centripetal* force arising only from an "attracting" *body* or central body[6] – that the vortex theory could not have led to mutual forces acting equally between two bodies. Since Newton was writing about vortices in relation to the moon's motion in the years after his alleged early discovery (*c.* 1666) of universal gravity acting according to the inverse-square law, we have yet additional evidence against the traditional story that Newton had discovered universal gravity in the 1660's and that there was a twenty-year delay (1666/7–1686/7) in his publishing the law.[7]

As to the aether itself, there were many vicissitudes in Newton's changing views. R. S. Westfall has chronicled Newton's early opinions on the aether, and Henry Guerlac and others have called attention to the differences between the aether that appears in Newton's writings in the 1670's and the later aethereal medium of the closing Queries of the *Opticks.*[8] In particular, Guerlac has stressed the fact that the aether is discussed primarily in Queries 17–24 of the second English edition of the *Opticks* (1717/1718), which were first published in that edition and were not included in the earlier Latin edition of 1706; he has also documented the role of Hauksbee's experiments in the later thought of Newton concerning electricity in relation to the aether.[9]

But even though the aether does not appear in a notable way in the *Principia* in 1687, and not in the *Opticks* (1704 in English, 1706 in Latin), we should not assume that Newton turned this concept on and off and on again, as if it were water flowing from a faucet. There are not only visible traces of the aether in the *Principia,*[10] but soon after that book was published, he wrote to Leibniz that 'some exceedingly subtle matter seems to fill the heavens' ('At caelos materia aliqua subtilis nimis implere videtur').[11] It had to be 'exceedingly subtle' because of the pendulum experiments,[12] which had set a limiting upper bound to the possible resistance an aether could offer to motion of a body through it.[13] In 1693 Newton even went so far as to write to Leibniz: 'But if, meanwhile, someone explains gravity along with all its laws by the action of some subtle matter, I shall be far from objecting.'[14] And in fact he embraced with alacrity and enthusiasm an attempt by Fatio de Duillier to account for gravity by an aether hypothesis, saying that this was the only possible 'mechanical' explanation of gravity.[15]

What is of greatest interest in the present context, however, is that in the late 1660's or early 1670's, Newton's studies of the moon's motion were founded on a supposed vortical motion in the Cartesian manner;[16] and that in the mid-1670's and even in the 1680's, on the eve of writing the *Principia*, he was still following much the same line. In the earlier years, his scheme explained certain features of the moon's motion by what D. T. Whiteside has called 'the pressure of the solar vortex upon the terrestrial one in which the moon rides in its "planetary" path'.[17] In the later recurrence of a vortical theory of the moon's motion, appearing in a draft of a reply to a letter written by Flamsteed (17 March 1681 N.S.), Newton took up the question of the sun's heat. He argued that 'the central parts' of 'the liquid matter swimming in the Sun' must 'become as hot as if the hot fluid matter surrounding it equalled the whole Vortex'. Newton concluded that the 'whole body of the Sun therefore must be red hot & consequently voyd of magnetism'.[18] We are here more interested in Newton's unquestioning echoing of Flamsteed's version of the vortex theory[19] than in his conclusion concerning heat in relation to the sun's lack of magnetism.

We have seen[20] that Newton's concept of *mutual* gravitation could not have been formulated any earlier than the beginning of 1685 (or, possibly, in December 1684). We may assume that his belief in a dense "alchemical" aether would have come to an end by that time.[21] I believe that it was at this time that he made the pendulum experiments reported in the *Principia*, specifically to make certain that the aether could not possibly be dense enough to produce the effects of gravitation. Because this was merely an experiment to confirm a belief already arrived at, and not an experiment of discovery to establish a new truth, we may understand why he did not bother to repeat the experiment at the time of writing the *Principia*, when he said that he could not find the sheet or sheets on which he had originally recorded his data and calculations. In the new scholium to the tract *De Motu* (originally written in Nov. 1684), Newton referred to 'the motions of the celestial bodies in the aether' (scholium following prob. 5) as an example of 'the motions of bodies in resisting mediums'.[22] He then explained why he thought that 'the resistance of pure aether is either nil or extremely small'. It is to be noted, however, that Newton does not then refer to the pendulum experiment, as he almost certainly would have done if he had already made such an experiment. Rather, he invoked a very poor example, that 'sailors in the open seas when protected from the winds feel nothing at all of the continuous flow of the aether'.[23]

The pendulum experiment gave Newton confirming evidence that the upper limit to the aether's possible density was too small to produce gravitation (or the apparent attraction of one body toward another), so that he could go beyond stages one and two of the "Newtonian style" and assume that such action occurs in the "system of the world" – although it was by no means clear as to how this "attraction" might be caused or might occur. But he was now completely free to explore the consequences of such a mutual gravitational attraction, a concept that the aethereal explanation of planetary motion effectively would have prohibited. There is no warrant for any assumption that an aether derived from alchemical considerations would (or even could) have led Newton toward the concept of *universal* gravity. The traces or vestiges of Newton's belief in the aether at the time of the *Principia* point to an aether of small density, one offering practically no resistance to the motion of matter through it, a non-alchemical aether.

It should be kept in mind that even though the aether does not appear to have exerted any creative influence on the *Principia*, Newton continued to cling to the belief that the aether (or an aether, or an aethereal medium or substance) might provide a "mechanical" explanation of gravitational force. This hope appears in the final scholium to sect. 11, Bk. 1, of the *Principia* and echoes of it may be found in his letters to Bentley and to Leibniz soon after the *Principia* was published.[24] It was Newton's yearning for such a "mechanical" explanation that drew him to embrace Fatio's hypothesis of an 'aether or such like subtile matter', which had the additional virtue that from its 'rectilinear motion in all directions, there is deduced the action of gravity in reciprocal proportion of the squares of the distances'. Newton's desire for an explanation of gravity that would, by way of the aether, satisfy the canons of the mechanical philosophy was so strong that it outlasted some forty years of failure to produce an aethereal demonstration of a force-action that would be mutual and that would be proportional to the masses and inversely proportional to the square of the distances. Universal gravity was apparently conceived without the support of an aether theory and in spite of the absence of such a theory. Hence, however much Newton may have cherished the illusion that the force of universal gravity could be reduced to the action of an aethereal matter and motion, the actual contribution of this point of view to the formulation of the principle of universal gravitation and the elaboration of the properties and actions of the force of universal gravity must be adjudged to have been nil.

How could Newton have been solving problems of orbital motion in terms of direct centripetal forces if, during the years 1679/80 to 1681, he was still writing about aethereal vortices? This question is not unrelated to another one: how could he have been investigating the action of solar planetary forces in 1679/80 and again in 1684, and how could he have been pursuing his studies of mutual forces of attraction (leading up to universal gravity) while writing the *Principia* during the years 1685–1687, if he could not bring himself to believe in forces acting at a distance? The answer is that the "Newtonian style" permitted him to explore the consequences of assuming a force to be centripetal or to act at a distance, without requiring any adherence to a belief in the real existence of such forces. Then, according to the "Newtonian style", after he had found that such forces "exist" (to the extent that the assumption that they exist enabled him, in Book 3 of the *Principia*, to retrodict and predict the known phenomena and further phenomena that tested the assumption), the problem was to reconcile such forces with the canons of the received philosophy of nature, to explain both how such forces could exist and how they could act according to the laws set forth and used in the *Principia*.

8. CONCLUSION

Among the topics which I have not discussed here, and which are directly related to the state of our knowledge concerning the *Principia* and its concepts, there is the question about whether the formation of Newton's ideas concerning universal gravity was influenced (either directly or indirectly) by his alchemical researches and his studies of the Hermetic philosophy. Other such topics are the Newtonian concept of mass, Newton's notion of continuous and instantaneous forces, and his views on space and time. Yet other questions relating to the *Principia* are the degree to which Newton actually employed inductive procedures (say, in establishing that gravity is universal), his possible use of the methods of analysis and synthesis (or resolution and composition), and the relation of Newton's theory of matter and particulate forces to his concept of forces (such as universal gravity) associated with gross bodies. Each of these would require a full-length essay, and here I shall merely indicate what some of the problems are, as I see them, in relation to the Newtonian universal gravity and to the "Newtonian style".

With regard to Newton's early statements regarding particulate forces and the forces associated with gross bodies, I have always been impressed by Newton's diffidence. In most of his writings, he seems to have been expressing hopes or beliefs about particulate forces, his aims or expectations, rather than making assertions about accomplished facts or guaranteed truths.[1] There was a good reason for adopting this position, since Newton had no real experimental or phenomenological evidence for the existence of such short-range particulate forces, nor had he ever constructed a successful mathematical system encompassing such forces in the "Newtonian style". Indeed, a characteristic aspect of Newton's writings on interparticulate forces and on the structure of matter is the general absence of mathematics in any true sense.[2]

Additionally, the received philosophy (or the "mechanical" philosophy), as I have said earlier, had no place for forces except as they arose from contact, as in the impact of a moving body on a second body. Hence, to introduce forces in the primary way that Newton did was to go against the general opinion on what was acceptable in natural philosophy and what was not. Phase three of the Newtonian style, displayed in Book Three of the *Principia*, in Newton's development of the system of the world on the basis of the mathematical principles of Books One and Two, gave Newton evidence for the existence of a "real force" of universal gravity and the quantitative law of its action. In the Preface to the *Principia* (first edition, 1687), he quite naturally suggested that if the force of universal gravity "really" exists, then why should not short-range particulate forces also be presumed to exist? Summing up his philosophy, in the General Scholium (ed. 2, 1713), Newton insisted without any qualification that universal gravity 'really exists' ('revera existat') and that it 'really . . . acts according to the laws which we have set forth'. But in the Preface he says only that he has been led 'to suspect somewhat' ('multa me movent, ut nonnihil suspicer') that the 'other phenomena of nature . . . may possibly depend on certain forces by which the particles of bodies . . . either are impelled toward one another and cohere . . . or are repelled from one another . . .' That is, by the time of the second edition, he publicly asserts the existence of universal gravity even though he cannot understand it and cannot find a cause for its mode of action; but with respect to the particulate forces (which, he says, are like gravity at least to the extent that their 'causes [are] not yet known') he only voices a suspicion. The evidence for universal gravity was firm and, to his way of thinking, unassailable; but the evidence for particulate forces was, by contrast, very weak.

It has been suggested, however, that Newton may have stated the 'opposite' of the true relation, when he advanced an 'argument [that] supports the existence of short range forces on the analogy of gravity'.[3] Newton's ideas about particulate forces are developed in many MS documents, including the unpublished *Conclusio* to the first edition of the *Principia*, and are to be found in particular in drafts of the Queries of the *Opticks* and in the later published Queries.[4] And we know how radical Newton understood these ideas about forces to be because he said so *expressis verbis*. For example, after the publication of the *Principia* (1687), he wrote out, in English, a 'Prop. 18' which was evidently to appear in a proposed 'Conclusion' to Book Four of the *Opticks*. This 'Prop. 18' was evidently recognized at once by Newton to have been a "hypothesis", and so he crossed out the heading 'Prop. 18' and put in its place 'Hypoth. 2', which was to be part of a general 'Conclusion'.[5] The first paragraph of this hypothesis reads:

Hypoth. 2. As all the great motions in the world depend upon a certain kind of force (which in this earth we call gravity) whereby great bodies attract one another at great distances: so all the little motions in the world depend upon certain kinds of forces whereby minute bodies attract or dispell one another at little distances.

And this leads him to the following argument:

How the great bodies of the earth Sun moon & Planets gravitate towards one another what are the laws & quantities of their gravitating forces at all distances from them & how all the motions of those bodies are regulated by those their gravities I shewed in my Mathematical Principles of Philosophy to the satisfaction of my readers[6]: And if Nature be most simple & fully consonant to her self she observes the same method in regulating the motions of smaller bodies which she doth in regulating those of the greater. This principle of nature being very remote from the conceptions of Philosophers I forbore to describe it in that Book least I should be accounted an extravagant freak & so prejudice my Readers against all those things which were the main designe of the Book: but [7] & yet I hinted at it both in the Preface & in the book it self where I speak of the inflection of light & of the elastick power of the Air but the design of the book being secured by the approbation of Mathematicians, I had not [doubted *crossed out*] scrupled to propose this Principle in plane words. The truth of this Hypothesis I assert not, because I cannot prove it, but I think it very probable because a great part of the phenomena of nature do very easily flow from it which seem otherways inexplicable: such as are chymical solutions, precipitations, philtrations, . . . volatizations, fixations, rarefactions, condensations, unions, separations, fermentations, the cohesion, texture, fluidity and porosity of bodies, the rarity & elasticity of air, the reflexions & refraction of light, the rarity of air in glass pipes & ascention of water therein, the permiscibility of some bodies & impermiscibility of others, the conception & lastingnesse of heat, the emission & extinction of light, the generation & destruction of air, the nature of fire & flame, the springinesse or elasticity of hard bodies.

It is certainly clear from such a phrase as 'extravagant freak' (and: this 'principle of nature being very remote from the conceptions of Philosophers') how aware Newton was that his own position differed from those of his fellow natural philosophers.

If Newton, however, had come to believe in these short-range particulate forces before he conceived of universal gravity, then two fundamental questions arise. One of them is the actual date of Newton's conversion to the possible existence of particulate forces (which could very well have occurred long before the *Principia*); the other is the presumed influence of this new belief on Newton's invention of the concept of universal gravity. We do not have any way of knowing the thought processes that led Newton to the great intellectual leap forward from a sun-directed centripetal force acting on planets to a force of universal gravity. In early 1685 and in the subsequent months, he saw how equality of action and reaction led to forces between planets (as worked out in the revised scholium to *De Motu*) and eventually to forces between any gross bodies and their parts. The "Newtonian style" enabled Newton to consider the actions of gravitational forces without initial reference to the question of their actual existence in physical nature, and eventually (phase three) he was able to show that the system of the world could be constructed on the basis of such forces: If he had already become fixed in the conviction that particulate forces exist, then it would perhaps have been somewhat easier for him to suppose (if even only mathematically, at first) that large-scale forces may exist. But I know of no evidence that any such train of thought ever occurred, nor do I know of any document in which Newton shows himself to have been thinking of such centripetal forces of (or acting on) gross bodies before his correspondence with Hooke in 1679/80. And there is equally no trace that before he wrote the revised scholium to *De Motu* (in early 1685, or barely possibly in December 1684), he had ever come to conceive of attractive or centripetal forces acting mutually between gross bodies. In any event, Newton would not appear to have arrived at his concept of universal gravity by simple induction from, or by an extension of, the forces of particles to forces between all gross bodies, so much as by the imaginative consequences of the systematic application of the third law of motion to a posited centripetal force.

In the *Principia*, as a matter of fact, Newton does not limit himself to mere considerations of analogy and existence in treating the relations between particulate forces and the forces of bodies such particles compose. In his treatment of this problem, in sect. 12, bk. 1, he shows us his maximally creative spirit. Here he is able to prove by rigorous mathematics (and not by analogy) that in the 'two principal cases of attraction', gross bodies will exert 'centripetal forces [that] observe the same law of increase or decrease in receding from the center as the forces of the particles themselves'.[8] These two cases are the square of the distance inversely and the distance directly, the only two that occur in nature. The first refers to the attraction exerted on an external body or particle by a homogeneous sphere or a homogeneous spherical shell (or a sphere composed of concentric homogeneous shells), and the second refers to the attraction exerted by a homogeneous solid sphere on an internal particle.[9] That this should be true of the 'two principal cases of attraction' is, as Newton says, 'very remarkable'. Here we see how careful and exact mathematical reasoning demonstrates the concurrence of properties of the two kinds of forces. This is in no sense an argument from analogy.

In Book 1, the mass or quantity of matter of bodies is not generally of major consideration, since Newton is dealing for the most part with mass points or unit masses and what he calls 'the accelerative measure' of forces,[10] that is, how much acceleration a force produces or would produce on a unit mass.[11] In Book 3, on the system of the world, however, he could no longer avoid the physics of real bodies. In prop. 8, he considers two mutually gravitating spheres (in which 'the matter in places on every side equidistant from the centers is similar') and proves that their 'weight' toward each other will be inversely as the square of the distance between their centers. This is what he says:

> After I had found that the force of gravity towards a whole planet did arise from and was compounded of the forces of gravity towards all its parts, and towards each individual part was as the inverse square of the distance from that part, I was yet in doubt as to whether the inverse-square proportion did accurately hold, or but nearly so, in the total force compounded of so many partial ones. For it might possibly be that the proportion which is exact enough at greater distances might be far from the truth near the surface of the planet, where the distances from the particles are unequal and their relative situation dissimilar. By the help of props. 75 and 76, bk. 1, and their corollaries, I was at last satisfied of the truth of the proposition as it lies before us.[12]

In this example we may, I believe, see how Newton would ideally proceed by strict mathematical proof to associate mathematically specified particulate forces with the more apparent ones of gross bodies. Accordingly, I would see the *Principia* and the dynamics of gross bodies leading Newton toward a wholly new kind of exact or mathematical theory of matter. This aspect of Newton's exact science is independent of, or at least supplementary to, the question of the creative influence of prior considerations of short-range particulate forces on Newton's concept of universal gravity and the science of the *Principia*.

With regard to analysis and synthesis, there arises a special problem in relation to the *Principia*. Most discussions of this topic do not bring in the *Principia* at all, but concentrate on various unpublished documents and the famous statement in Query 31 of the *Opticks*, in which Newton sets forth the way in which 'in Natural Philosophy, the Investigation of difficult Things by the Method of Analysis, ought ever to precede the Method of Composition'. What Newton then does, in the penultimate pages of the *Opticks* (in the later editions), is to give a sound exposition of the classical methods of investigation in experimental science, which he there calls 'experimental Philosophy'.[13] What is confusing, however, is that this method is misleadingly introduced by the phrase: 'As in Mathematicks, so in Natural Philosophy'. But in mathematics, and hence in the *Principia*, the use of the words "analysis" and "synthesis" may be exactly the reverse of what it is for experimental science. This may be seen most clearly by comparing what Newton himself gives as examples of 'analysis' and of 'composition' (or of 'synthesis') in the *Opticks* and in the *Principia*. This aspect of Newton's philosophy was fully understood a century and a half ago by the Scottish philosopher Dugald Stewart, who gave examples to show the imprecision of Newton and the Newtonians in apparently relating the modes of investigation in physical or natural philosophy and in mathematics. Stewart even found that in some cases 'analysis' and 'synthesis' had opposing senses in the two realms.[14]

The rôle of induction in relation to universal gravity needs careful specification. Let us review some of the main steps (in Book Three of the *Principia*) leading to the establishment of universal gravity. Newton begins with a primary assumption concerning centripetal forces. For the satellites of Jupiter (prop. 1, bk. 3), the area-law (hypoth. 1, later phen. 1, bk. 3) shows (by props. 1 & 2, bk. 1) that there is a force directed toward Jupiter; Kepler's third law indicates (prop. 4, bk. 1) that this is an inverse-square force. Similar considerations apply to the motion of our moon[15] (prop. 3, bk. 3). Then (prop. 4 & schol., bk. 3) Newton shows that the moon "falls" constantly toward the earth (that is, "falls" away from its inertial linear tangential path) with the same rate of fall toward the earth that any body would have at that distance (60 earth-radii) from the earth's center, assuming that the earth's gravity could extend that far, diminishing according to the law of the inverse square. In other words, Newton assumes that at 60 earth-radii from the earth's center a body would – if the earth's gravity extends out as far as the moon and varies inversely as the square of the distance[16] – fall toward the earth with a rate 1/3600th of what is observed to be the case at 1 earth-radius from the center (or at the earth's surface). When calculation verifies the argument, he says that the force acting on the moon is from then on to be called gravity.[17] The situation proves to be the same, of course, for any planet with a satellite and for the planets with respect to the sun (whose satellites they may be considered to be). All are subject to the same inverse-square law of force. And all such satellites, by the argument from the third law of motion (as we have seen in §5 *supra*), and their respective primaries, are both sources of attractive force and bodies on which attractive force acts, and so act on one another. And now comes an argument from induction: not an induction from a property found in all bodies that have been observed to a property of all observable bodies, or from such a property of all bodies that can be observed to bodies that cannot be observed, but rather from bodies that have a satellite or a system of satellites (sun, earth, Jupiter, Saturn) to those that do not (e.g., Mercury, Venus, Mars).[18] The induction is not related to an observed property (e.g., iron is hard, water flows, Socrates is mortal), but rather to the result of a mathematical derivation concerning the quantitative statement of a law of attraction. In the end of the sequence of Newton's argument, it is the distinguishing property of "mass" in all samples of matter or bodies that makes all matter or all bodies gravitate toward all other matter or bodies.

I have referred to the suggestion made by R. S. Westfall, that Newton had a radical change in his thinking about forces. Presented in Westfall's study of *Force in Newton's Physics*, this is stated to have been nothing less than a 'redirection of his [Newton's] philosophy of nature' – 'undertaken . . . about 1679'. The 'significance' of this radical shift can, in Westfall's words, 'scarcely be overstated'.[19] In dramatic and unambiguous language, Westfall described this 'redirection' as follows:

> What he [Newton] proposed was an addition to the ontology of nature. Where the orthodox mechanical philosophy of seventeenth-century science insisted that physical reality consists solely of material particles in motion, characterised by size, shape, and solidity alone, Newton now added forces of attraction and repulsion, considered as properties of such particles, to the catalogue of nature's ontology. As I shall argue later, the ultimate ontological status of forces in Newton's conception of nature is a complex and involved question. In his published writings, he chose to refer to his true opinion, as I understand it, only obliquely, and we must consult unpublished manuscripts to seize his meaning. As far as the works he published are concerned, and above all, as far as the conceptual tools he employed in scientific discussion are concerned, he treated forces as entities that really exist. In orthodox mechanical philosophy, the moving particle was the ultimate term of explanation. In confronting magnetic and static electric phenomena, for example, the kinetic mechanical philosopher refused to employ terms such as 'attraction', which were held to be occult, and turned in preference to the imaginary construction of invisible mechanisms whereby particles of matter produce the motions observed. Newton's philosophy of nature did not cease to be mechanical. Physical reality still consisted of material particles in motion, but the ultimate term of explanation was now the force of attraction or repulsion that altered a particle's state of motion. Newton's conception of nature has been called a dynamic mechanical philosophy.[20]

It was the new ontological status of force, according to this view, that would have enabled Newton, after the correspondence with Hooke in 1679/80, to solve the problem of motion according to Kepler's laws and eventually to generalize the concept of a solar-centripetal force into a force of universal gravitation, or gravitational attraction, on which the *Principia* is built.

In *Newton and the Concept of Force*, Westfall did not attempt to root Newton's concept of force or of universal gravity in Hermeticism, although he did say of Roberval that *his* 'conception of gravity, which was virtually identical to that given by Kepler sixty years earlier . . . is indistinguishable from the sympathies of Hermetic natural philosophy'.[21] Four years later, however, in an essay on 'The Role of Alchemy in Newton's Career', Westfall argued that the source of Newtonian gravitation was to be found in alchemy, in a transformation of an alchemical 'active principle'[22]: 'The essence of an attraction in contrast to an active principle was its quantitative definition, whereby it might fit into the structure of rational mechanics.'[23] In commenting on Newton's argument in the *Conclusio*[24] that Newton wrote for, but did not publish in, the first edition of the *Principia*, Westfall summarized Newton's argument for short-range forces between particles: '. . . that nature is exceedingly simple and conforming to herself'. Thus, 'Whatever reasoning holds for greater motions, should hold for lesser ones as well'. According to Newton, 'The former depend upon the greater attractive force of larger bodies, and I suspect that the latter depend upon the lesser forces, as yet unobserved, of insensible particles'. Westfall's comment then reads:

> The structure of the argument supports the existence of short range forces on the analogy of gravity. When we recall where Newton's attention was fixed during the decade before 1685, we are apt to conclude that the relation of the two was exactly the opposite. The forces of attraction and repulsion between particles of matter, including gravitational attraction which was probably the last one to appear, were primarily the offspring of alchemical active principles.[25]

Westfall's conclusions have influenced B. J. T. Dobbs's presentation[26] of the ways in which Newton made an 'integration of alchemy and mechanism', one of which was Newton's 'inclusion of attractive forces, the "active Principles" of the Hermetic tradition which Westfall has also seen incorporated into the attractive forces of gravity'.[27] In particular, she says that she follows Westfall in describing Newton's work 'on the dynamics of orbital motion' of 1679/80 as an application of Newton's 'chemical idea of attraction to the cosmos', proposing that 'the concept of alchemical attraction had allowed him [Newton] to accept the ontological reality of celestial forces'.[28] One feature that tends to make this point of view attractive to scholars is that it offers a plausible source for Newton's introduction of a new mode of accounting for phenomena by the action of forces:[29] going beyond the then-accepted canons of explanation, such as the action of matter and motion, the two fundamentals of the mechanical philosophy.

I do not wish to enter here into the question of whether there was such a redirection of Newton's philosophy of nature, on which the alchemical interpretation of attractive forces depends. I know of no document of the period 1675–1686 – that is, prior to the *Principia* – which enables us to give either a simple and direct proof or a rebuttal of the proposal that at some particular time within this interval, there had been a distinct change in the ontological status given to force by Newton.[30] But there can be no doubt that by the time that Newton was writing the unpublished *Conclusio* to the first edition of the *Principia*, he did (however tentative the forms of expression) show that he had come to believe in the real possibility – if not the actual reality itself – of inter-particle forces acting without the mediation of anything else (such as an aether).

It is an undeniable fact, moreover, that soon after (or perhaps even during) Newton's correspondence with Hooke, and certainly within a few years after that correspondence, Newton was indeed exploring the consequence of forces that would directly pull or push one gross body toward another: for example, the planets toward the sun. But whether or not there was actually a concomitant complete change of Newton's philosophy, so that thereafter he was willing to give force the same ontological status as matter and motion, the "Newtonian style" which he developed *did* enable him to consider problems of motion associated with the heavenly bodies *as if* these bodies were being pushed or pulled by forces, whether or not he then began to conceive of 'forces as entities that really exist'.[31] We have seen that from the point of view of the "Newtonian style", such investigations (phases one and two) could be – and actually were – undertaken without the question ever arising (at that stage) of Newton's commitment to the independent existence or reality of such forces. When Hooke suggested to Newton that orbital motions be analyzed in terms of a linear inertial tangential motion and a force directed toward the sun, Newton not only could and did proceed with his analysis, without having to make an absolute commitment as to whether *there are* such forces; he did not suffer any inhibition from a premature concern as to whether *there could be* such forces. We have seen how such inhibitions prevented Huygens from making similar advances in celestial dynamics well within his potential.[32] I am not arguing here that Newton did not have the 'redirection' of his 'philosophy of nature' that has been suggested, but rather that, in view of the "Newtonian style", such a redirection was not an absolute necessity for Newton's being able to deal with forces as he did in his work in celestial dynamics culminating in the *Principia*.

A key to Newton's thinking about macroscopic forces at the time of the *Principia* is to be found in that famous sect. 11 of Book 1 of the *Principia*, where he introduced the term "attraction" in a mathematical sense.[33] As we have seen, he made it clear to any reader that what he called "attractions", as a mathematical name for mutual 'centripetal forces', would 'in the language of physics ... more truly be called impulses'. His idea, that bodies are "drawn" toward one another *as if* by mutual centripetal forces, that they are not physically "attracted" but rather impelled, occurs not only in the introduction to sect. 11, but again – as we have seen – in the conclusion to sect. 11. Attraction, he said there, is used 'in a general sense' for any process whatever by which bodies 'endeavour ... to approach one another'. This could be the result of 'the action of the bodies either mutually seeking one another or agitating one another by means of spirits emitted' (as in the case of electric effluvia[34]); or it could arise 'from the action of aether or of air or of any medium whatsoever – whether corporeal or incorporeal – in any way impelling toward one another the bodies floating therein'.[35] This passage shows us a Newton who seems not to have believed in a force of attraction as being a primary quality so much as being a secondary effect of something else: a result of the action of some other entity that was more philosophically acceptable to himself as well as to his fellow natural philosophers.[36] Thus Newton believed that he had shown in the *Principia* that gravity really exists and acts according to the law which he had found. But for Newton that meant only that (as a sequence to phase three of the "Newtonian style") he now had to reduce such an "unacceptable" force to some other "acceptable" action. Newton's efforts to find some such reasonable explanation of gravity have been chronicled for us by R. S. Westfall.[37] They included some kind of 'impulses', electricity and the action of a 'subtle spirit', the action of an elastic 'aethereal medium', apparently one that was not a material aether. All of these efforts were directed toward a single end, expressed in his letter to Richard Bentley in 1691, when he said that it was 'unconceivable that inanimate brute matter should (without the mediation of something else which is not material) operate upon & affect other matter without mutual contact ...' Further, that 'gravity' is of such a nature that 'one body may act upon another at a distance through a vacuum without the mediation of anything else by & through which their action or force may be conveyed from one to another is to me so great an absurdity that I believe no man who has in philosophical matters any competent faculty of thinking can ever fall into it'. He concluded:

> Gravity must be caused by an agent acting constantly according to certain laws, but whether this agent be material or immaterial is a question I have left to the consideration of my readers.[38]

Presumably in despair that no reductionist "explanation" of gravity was forthcoming, he would – according to Fatio de Duillier – 'often seem to incline to think that Gravity had its Foundation only in the arbitrary Will of God'.[39] Toward the end of his life, he would turn again to the aether as he had done in his young manhood, although now the aether was of a different sort; but again the varying degrees of density were thought to account for gravity, this time universal and not terrestrial gravity.

Today, thanks primarily to the researches of R. S. Westfall, Betty Jo Teeter Dobbs and Karin Figala,[40] there can no longer be any question as to the reality of Newton's concern with alchemy and Hermetic subjects.[41] We now know how deeply Newton had read in the Hermetic philosophy and in the literature of alchemy and how committed he was to a program of experimental alchemy and chemistry. During the decades before the *Principia*, alchemical reading and research was a major part of his intellectual life.

Nor should this be surprising. Whoever reads the writings of Christopher Hill will know that concern for esoteric subjects such as alchemy and the Hermetic philosophy, prophecy, and astrology was very characteristic of the thinking men of that age.[42] Robert Boyle and John Locke were also deeply concerned with alchemy. Newton would have been a great exception had he been free of any contact with alchemy and the Hermetic philosophy. Newton differed from many of his contemporaries with respect to alchemy and prophecy, however, in that he had a disdain for astrology, with which alchemy was then often associated.

Newton's Hermetic initiation and experience must certainly have affected his language, concepts, and mode of thinking in other areas, consciously or unconsciously. It would be hard to believe that Newton's thinking about forces in the creative years from 1679 to 1687, during which the *Principia* was created, would not to some degree have been tinctured by the alchemical background. But this is very different from finding that a particular concept or method may have had a direct alchemical or Hermetic source. The new concept which got Newton going on the path to universal gravity and gravitational celestial mechanics was centripetal force, a replacement of the older Cartesian *conatus recedendi a centro* and the more recent Huygenian *vis centrifuga*. The source of Newtonian *vis centripeta* is not to be found in alchemy or the Hermetic philosophy, but occurs in the letters sent by Hooke to Newton in 1679 and 1680, stating clearly and unambiguously how to analyze orbital planetary motions.[43] Perhaps Newton's explorations of alchemy may have helped to condition him to be willing to accept or use the notion of a *vis centripeta*, may have made his mind receptive to Hooke's suggestion. But what seems to me of far greater import is that the "Newtonian style" enabled him to use and to develop the properties of *vis centripeta* without having fully to commit himself to this concept as a fundamental element in the philosophy of nature, one that did not need to be explained by some more acceptable ancillary concepts so as to become accommodated to the reigning mechanical philosophy. That is, he was able to explore the mathematical consequences of a celestial dynamics based on a *vis centripeta* (which quickly became a system of mutual centripetal forces of "attractions") without having at that stage to commit himself (or even to inquire) as to whether or not such forces do or can exist.

On the basis of accumulated scholarly evidence, we may agree that Newton's alchemical studies are related to his concept of a "universal spirit" and possibly to a 'distinction between nature's "vegetable" actions and the "purely mechanical ones" '.[44] It should always have been axiomatic that Newton's theory of matter, his notions concerning the interactions between varieties of matter, and his views on the properties and forces associated with corpuscles of matter, cannot be fully understood apart from their alchemical and Hermetic context. And it is the same with respect to Newton's concept of "active" (vs. "passive") principles. Additionally, various aspects of the aether, as propounded by Newton, were – at least in the 1670's and early 1680's – conditioned by alchemical notions and by Hermetic ideas; this is obvious to anyone who reads the hypothesis on the aether of 1675.[45] The recent research and speculation by R. S. Westfall, Betty Jo Teeter Dobbs, and Karin Figala provide a firm link between alchemy or Hermetic philosophy and the development of Newton's thinking in such areas as the corpuscular theory of matter, the structure and properties of matter, chemistry and interactions between different types of material corpuscles, the inter-particulate forces, "active" principles, speculations concerning electricity and magnetism, and some aspects of optics.[46] One may even find grounds for linking Newton's ideas about the cause of TERRESTRIAL *gravity* with his alchemical studies.[47] This is an impressive list, but it does not include the two main concepts that made the *Principia* great: mass and universal gravity.

At this point a certain amount of precision is needed. Gravity is the phenomenon in which a terrestrial body is pulled or pushed (impelled) toward the earth, or acts *as if it were*. This earthward propensity or "gravitas" is also called "weight". Early on, as I have mentioned, Newton thought to *explain* THIS gravity by a descending aethereal shower, as he later (in his letter to Boyle in February 1679) sought to *explain* gravity by supposing 'aether to consist of parts differing from one another in subtilty by indefinite degrees'; he also tried to *explain* gravity (in his hypothesis in 1675) by the possibility that 'the gravitating attraction of the earth [may] be caused by the continual condensation of some . . . aethereal spirit', arguing in part from the supposed aethereal nature of electrical attraction. In all three cases, he sought to *explain*, to find out *how* the 'gravitating attraction of the earth' might – as he said – 'be caused', to conjecture 'about the cause of gravity', the cause of terrestrial weight. This line of speculation presumably cannot be wholly divorced from an alchemical context, but – be it noted – it did not ever lead to an elucidation of any of the properties of gravity – although Newton later claimed that his aether hypothesis implied an inverse-square law.[48] This kind of speculation did not even lead Newton to the idea that *terrestrial gravity* extends to the moon. We have seen that Newton was then explaining the moon's motion by a Cartesian *conatus recedendi a centro* and not by a centripetal force, directed toward the earth as center. The clear and unambiguous notion of a lunar and planetary *centripetal force* came to him directly from Hooke in the course of their correspondence in 1679–1680. Only then did the question arise as to whether the *lunar centripetal force* arising in the application of dynamics to elliptical (or near-elliptical) orbits could be physically the same as terrestrial gravity. Alchemy and Hermeticism did not lead Newton to the idea that terrestrial gravity may extend as far out as the moon's orbit.

Universal gravity has both a qualitative and a quantitative aspect, of which we may limit the discussion here to the qualitative or the conceptual part.[49] *Terrestrial gravity* may seem to have a "universal" aspect in that *all* heavy bodies which we encounter in experience 'tend toward the Centre of the Earth'; but the more usual sense of UNIVERSAL *gravity* is that of *all matter* tending toward *all other matter*. This true "universality" cannot be disassociated from the mutuality of gravitation. But Newton radically altered the sense of the word "gravity" when he concluded that the earth "gravitates" or is "heavy" toward an apple to exactly the same quantitative degree that an apple "gravitates" or is "heavy" toward the earth. Gravity thus meant no longer only the force of an object toward the earth (as weight had implied) but also the simultaneous equal and opposite force of the earth toward an object. In Book 3 of the *Principia*, Newton proved that the force of the earth on the moon is the same as gravity and (prop. 5) that 'the force by which celestial bodies are kept in their orbits . . . is gravity', so that (scholium to prop. 5) 'the cause of the centripetal force by which the moon is kept in its orbit ought to be extended to all the planets'. This led him to discuss how 'all bodies gravitate toward each of the planets' and have 'weight toward any one planet' (prop. 6); and eventually (prop. 7), he reached the generalization of 'a power of gravity pertaining to all bodies' proportional to their respective masses. In physics, from that day onwards, "gravity" has no longer had the older restricted meaning of "terrestrial weight".

Unless it can be shown that alchemical considerations led Newton directly to the concept of third-law mutuality and played a rôle in the successive steps leading from terrestrial weight to universal gravity, it is difficult to see what direct contribution his alchemical studies or reading in the Hermetic philosophy may have made to UNIVERSAL *gravity*, however much (in the 1670's) alchemy and the aether may have been linked to attempts to "explain" (in the sense of assigning a "cause" to) TERRESTRIAL *gravity* or terrestrial weight. Alchemy or the Hermetic philosophy had not made the planetary centripetal (really solipetal) force acting on the planets a mutual force in the first version of *De Motu* in November 1684. We have seen that what led Newton to propose such a mutuality was – by his own testimony, and as the present analysis has shown – the application of the third law, that "action" must always be accompanied by an equal and opposite "reaction".

I have developed the features of the "Newtonian style" only in the context of the *Principia*, where they may be found most obviously displayed. The question must necessarily arise as to the "Newtonian style" in relation to Newton's researches in the theory of matter and chemistry (or "alchemical" chemistry) and optics. I have mentioned earlier that an outstanding feature in Newton's theory of matter and of chemical reactions is the absence of a full mathematical development: a lack that, no doubt, arose because of the relatively primitive state of knowledge in those subjects. The result is that Newton did not develop his theory of matter and of chemistry in the "Newtonian style" in which the science of the *Principia* is unfolded. As to optics, Newton certainly did attempt to produce systems in the "Newtonian style"; that is, he did seek to explore the consequences of a mental construct (or even a model) by means of mathematics. One of those was published in sect. 14 of Book 1 of the *Principia*, proposing a way of accounting for aspects of refraction by the action of centripetal forces on the motion of minute corpuscles.[50] Other such essays in the Newtonian style occur in Newton's MSS and in his *Lectiones Opticae*; their properties have been studied by Alan Shapiro,[51] J. A. Lohne,[52] and especially by Zev Bechler.[53] But they were, unlike the *Principia*, essentially failures. No one has yet made a thorough study of the extent to which there might have been a mathematical underpinning of the general conclusions in the *Opticks,*[54] but it is patently obvious that the formal presentation in that treatise is not in what I have here been calling the "Newtonian style". In the *Opticks*, Newton does not start out with a mental construct (or a model) and then deduce its consequence by the application of formal mathematical techniques: algebra, proportions, geometry, methods of infinite series, theory of limits, fluxions. Rather, he gives a 'Proof by Experiment'. In this sense, the *Opticks* is not a mathematical treatise,[55] and is certainly not an exercise in the "Newtonian style" of the *Principia.*[56] And the reason is that Newton's attempts to construct a science of optics in the Newtonian style were never successful.[57]

In tracing the roots of universal gravity, I have stressed the importance of the law of action and reaction and the crucial rôle of the "Newtonian style". I have stated my conclusion concerning the lack of convincing evidence that Newton's pursuit of alchemy or his studies of Hermetic philosophy directly influenced that line of his thinking in dynamics which led to the *Principia* and the Newtonian system of the world. But I am perfectly willing to admit the possibility that the development of a system of mechanics in the "Newtonian style", in which forces play a significant rôle, was affected[58] or reinforced by considerations of particulate forces, as developed in relation to an alchemical or Hermetic background. What seems to me remarkable about Newton in this regard, however, is that although he was steeped so deeply in alchemy and the Hermetic philosophy, his *Principia* is so little tinctured by this thought and activity. Even if it should turn out that there is some secret and hitherto-unrevealed direct effect of Hermeticism on the *Principia* – then it would still be remarkable that there are no great apparent or obviously visible traces of alchemy or Hermeticism in the *Principia.*[59] Such traces, if any, would have been hidden so well that they would have remained indiscernible to scholars and critics for three centuries. In short, what may be most astonishing about the *Principia* is not that its author was deeply touched by alchemy and the Hermetic philosophy, but that he should have been able to put aside his alchemical and Hermetic concerns to such an extent that he could, despite them, produce a work of positive science which has influenced the development of the exact sciences and of all science ever since. This is in itself remarkable, a wholly unexpected kind of behavior. But then – as Frances Yates once remarked to me – in dealing with Newton we must always remember that he was a genius; a man with the mind of a Newton does not necessarily follow the expected rules for other human beings.

Harvard University

NOTES

*This essay is a somewhat revised and expanded version of a talk given at the Newton Symposium, held in Churchill College, Cambridge, 7–9 August 1977, in commemoration of the 250th anniversary of Newton's death.

The research on which this paper is based was supported by a grant from the National Science Foundation (U.S.A.).

§ 1: INTRODUCTION

[1] The earliest reference I have found to a revolution in the sciences (in the sense in which we commonly use the word "revolution" today) occurs in 1754, in an essay by Clairaut, in which it is said: 'Le fameux livre des Principes mathématiques de la Philosophie naturelle (de Newton) a été l'époque d'une grande révolution dans la Physique'. Earlier, Bernard le Bouyer (*or* Bovier) de Fontenelle referred to the calculus, which he associated with Newton and Leibniz as its co-discoverers or its co-inventors, as a 'révolution presque totale ... dans la géométrie'. On Fontenelle and Clairaut, see Cohen (1976*a*), and Cohen (1982).

[2] *I.e.*, Cuvier, van 't Hoff and Ostwald; see Cohen (1980), p. 293, n. 19.

[3] The same is true of most classics in the exact sciences: Ptolemy's *Almagest*, Copernicus's *De revolutionibus*, Kepler's *Astronomia nova*, Galileo's *Discorsi*, Laplace's *Mécanique céleste*. The opinion is often expressed that the main difficulty in reading Newton's *Principia* lies in the fact that it is written in an archaic geometric style, that it is not cast in the language of the calculus. In fact, the difficulty rather lies in Newton's condensed presentation, in his often imperfect and incorrect procedures and conclusions, and in the inherent complexity of the subject-matter.

[4] Introduction to *Principia*, Book 3. All English versions quoted in this article are taken from a new translation of the *Principia* by I. B. Cohen and Anne Whitman.

[5] Newton's preface to the first edition is a crucial document. Cotes's lengthy preface to the second edition is perhaps of more importance as a key to understanding the issues raised in debates concerning the Newtonian philosophy in general than as an aid to reading the *Principia*. Newton's prefaces to the second and third editions deal only with matters of detail. The preface to the first edition should be read in conjunction with an unpublished MS version and an unpublished MS conclusion, both first printed in Hall & Hall (1962), pp. 302–308, 320–347.

[6] Newton himself, in the introduction to Book 3, said: 'It will be sufficient to read carefully the Definitions, the Laws of Motion, and the first three sections of Book 1, ... consulting at will the other propositions of Books 1 and 2 which are referred to here [in this Book 3]'.

7 There is a preliminary version of what became Book 3 of the *Principia*, written by Newton before he decided to cast it in a mathematical form; it is often known as *Newton's System of the World*. For the relation between this work and Book 3 of the *Principia*, see Cohen (1971), suppl. 6.
8 These 'Phaenomena' and 'Regulae philosophandi' were grouped together as 'Hypotheses' in the first edition of the *Principia*.
9 On the "Newtonian style", see §6 *infra*.
10 On this alternative approach to celestial dynamics (prop. 41 vs. props. 1–14) and the reactions to it by Continental mathematicians, see Aiton (1964) and D. T. Whiteside's commentary on pp. 345–356 of vol. 6 of his edition of Newton's mathematical papers.
11 Quoted from Ball (1893), pp. 88–89, where there may be found a succinct but comprehensive account of the contents of the corollaries.

§2: SOME SOURCES OF INACCURACIES AND MISUNDERSTANDINGS CONCERNING THE *PRINCIPIA*

1 This is part of a longer autobiographical statement, of which a major portion was first published in *A Cataloque of the Portsmouth Collection* (1888), p. xviii; Newton's statement was printed again by Rouse Ball (1893), p. 7, who observed that the 'original is cancelled, and hence it is not indisputable evidence'; nevertheless, he said, that 'it is believed, that, except for certain dates in the paragraph next below, it is substantially correct'. The complete text was first published in Cohen (1971), suppl. 1; see the discussion in Newton's *Correspondence*, vol. 6, pp. 457–462.
2 This 'delay' was given special prominence in Cajori (1928). Cajori expounded a theory that the 'twenty years' delay' was caused by Newton's inability to prove the theorem that a uniform sphere, a uniform spherical shell, and a sphere composed of a series of concentric uniform spherical shells will act gravitationally as if the total mass were concentrated at the geometric center; Cajori was only developing at large a statement in an address by J. L. Glaisher on the 200th anniversary of the *Principia* in 1877, a thesis which Glaisher said had originated with J. C. Adams. On this topic, see my article in the introduction to the proceedings of the Newton meetings (Aug. 1977) held in Cagliari, published in the series *Vistas in Astronomy* vol. 20 (1979). The alleged 'twenty years' delay' is discussed later on in the present article.
3 In fact, Newton had at first written 'I having deduced that the forces which keep the planets in their Orbs'; then he changed the 'I having' to 'I', and started to insert 'centripetal' before 'forces' and cancelled it before writing the final 'I'.
4 Newton did not actually make the shift to centripetal forces before his correspondence with Hooke in 1679–1680. On the development of Newton's concept of centrifugal forces and the transition to centripetal forces, see Westfall (1971).
5 On Newton's 'conatus', see Westfall (1971), Whiteside (1970), and Herivel (1965).
6 See Whiteside (1970), p. 11: 'In exact terms, what Newton found in 1666 (or a little before or a little after – this is not pertinent here) was a "law" deducing that the *centrifugal* "endeavours" (*outwards* from the centre of motion) of planets travelling in pristine *circular* paths (wherein they were borne by the deferent solar vertex) varied as the inverse-square of their distance from the Sun; and similarly, *mutatis mutandis*, for other objects – in particular the Moon – caught up in the terrestrial whirlpool'. Furthermore, the very name of the law of universal gravitation implies that what is at issue is a *mutual*

gravitating force acting equally on the attracted and the attracting body. On the mutuality of the gravitating force, see § §6, 8 *infra*; also Cohen (1976) and Wilson (1970).

[7] The opening paragraph refers to the reason why he 'did not publish' his book *De quadratura* 'sooner', and then goes on to mention 'the testimony of these two ancient, knowing & credible witnesses'. The latter are Isaac Barrow and John Collins; the context is the preparation of the documentary evidence to support Newton's priority in the discovery of the calculus (for the *Commercium epistolicum*). See Whiteside (1966), p. 38, n. 3.

[8] The *Catalogue of the Portsmouth Collection* (1888) records (p. xix) that 'Newton appears to have made one or two mistakes of date, and probably for this reason has drawn his pen through the entire passage'.

[9] [Newton] (1715), pp. 173–224; see p. 206. For similar statements see Cohen (1971), suppl. 1, e.g., p. 294.

[10] Cf. Whiteside (1970), p. 119: 'The published state of the *Principia* – one in which the geometrical limit increment of a variable line-segment plays a fundamental rôle – is *exactly* that in which it is written'.

It is a curious fact that there exists for the *Principia* no extended set of primary working papers or early drafts in Newton's hand. It is as if he had systematically gone through his tremendous collection of MSS in order to weed out and destroy any evidence of the way he had actually 'found out . . . the Propositions in his *Principia Philosophiae*'. On this topic, see Cohen (1971), ch. 3 & 4.

[11] See Baily (1835), p. 690. Of the general term "theory" (and in particular "lunar theory"), Baily says: 'In Flamsteed's time its meaning in astronomy rather denoted rules or formulae for constructing diagrams and *tables* that would represent the celestial motions and observations with accuracy, than the development of the *causes* of phenomena, which is its rigid and more modern acceptation. But, in this latter sense, it could not properly be used prior to the appearance of the *Principia*, since those causes were not then known'.

[12] On this topic, see Whiteside (1976), Waff (1976); also Waff (1975) and Chandler (1975).

[13] In the *Acta* for March 1714, p. 140, the anonymous reviewer wrote: 'Indeed, the computation made of the lunar motions from their own causes, by using the theory of gravity, the phenomena being in accord, proves the divine force of intellect and the outstanding sagacity of the discoverer.' See Cohen (1975), p. 41.

[14] Another misunderstanding arises from the failure to take cognizance of the way a creative mind develops; often statements of Newton made as late as 1713 will be taken to explain a point of view held as early as 1672– as if there had been no change during some forty years.

[15] See §8 *infra.*

[16] The version 'I do not feign hypotheses' was first proposed in Koyré (1955) and Koyré (1956). It was later shown in Cohen (1962) that Newton did in fact write in English that he did 'not feign hypotheses'; he also said that he did 'not frame hypotheses'. Contrary to what many writers have imagined, however, in Newton's day the word "frame" was apt to have as pejorative a sense as "feign".

[17] Mach (1960), p. 236: ch. 2, pt. 3, §6.

[18] On the planned conclusion written by Newton for the first edition, see note 5 to §1 *supra.*

19 See Koyré (1955); Cohen (1956), pp. 130–134; and Cohen (1966).

20 I refer here to the mature Newton, not the youthful Newton, trying to understand and to systematize the concepts he had been encountering in his formal studies and his early reading. Concerning the essays of this sort, written out by Newton in various notebooks and in the 'Waste Book', see Westfall (1971), ch. 7, Herivel (1965), pt. 2, §§1, 2. An exception to my general statement may be Newton's philosophical critique of Descartes's principles in the essay beginning 'De gravitatione et aequipondio fluidorum . . .', published in Hall & Hall (1962), pp. 89–156, in which Newton does approach the concepts of dynamics in the context of abstract thinking or philosophical analysis rather than in the context of the mathematical solution of problems posed by physics. But Newton does not fully develop any of his major concepts in this youthful exercise.

21 Manuel (1968), p. 84.

22 The main assignment for the Cambridge symposium on Newton was to indicate areas in which research is needed.

23 It is important to have "influences" in the plural. The influence could be in terms of method, result, an inadequacy that called for research, an error to be corrected, a concept to be used or transformed, and so on.

24 It would be an error to assume that all these heroic figures were strict Newtonians, or that Newton had actually done more than open the door or show the way. He had not, in fact, even begun to deal with many major topics in mechanics that remained to be explored in the Age of Reason. See Truesdell (1960) and Truesdell (1970).

25 Even Darwin referred to Newton when seeking a model for good science.

26 This point is developed in Cohen (1980), Cohen (1981), and Cohen (1982).

§3: NEWTON'S ANALYSIS OF THE MOTION OF THE MOON IN THE 1660's

1 For the literature concerning Newton, see the bibliographies in Cohen (1971), Westfall (1971); also Cohen (1960), Whiteside (1962). Pighetti (1960) and Whitrow (1971) give keys to the literature. See also Gray (1907); Babson Collection (1950); Wallis & Wallis (1977).

2 On Newton and Descartes, see Koyré (1965), pp. 53–200 ('Newton and Descartes'), Herivel (1965), pt. 1, §2 ('The influence of Galileo and Descartes on Newton's dynamics'). Whiteside (1970), pp. 10sqq., Cohen (1964); on Newton and Galileo, see Cohen (1967).

3 See §2 *supra*, and the latter part of this §3.

4 This was disclosed with the publication of Stukeley (1936); interpreted by Pelsneer (1937) and Cohen (1946). See McKie & De Beer (1952).

5 See Herivel (1960*a*); and Herivel (1965), pp. 7sqq., 128sqq.

6 In the final paragraph of the scholium following prop. 4, Bk. 1, of the *Principia*, Newton gives a résumé of his early derivation of the rule for centrifugal *conatus* or centripetal force, but he does not say that this represents an early discovery made by him in the 1660's. This paragraph was not part of the original MS sent by Newton to Halley. But when Halley reported that Hooke wanted Newton to give him credit for the inverse-square law, Newton sent this paragraph on to Halley with the following explanation: 'In turning over some old papers I met with another demonstration of that proposition, which I have added at the end of this Scholium.' If he knew the v^2/r rule and was

acquainted with Kepler's third law, he obviously could have combined them to get an inverse-square law.

7 See the autobiographical note quoted at the beginning of § 2 *supra.*

8 But *not* an inverse-square law of CENTRIPETAL *force.*

9 See Ball (1893), p. 13.

10 U.L.C. MS Add. 4004, fol. 1; transcribed and analyzed in Herivel (1960) and Herivel (1965), pp. 7–11, 129–132.

11 This occurs in the final paragraph of the scholium to prop. 4, Bk. 1, of the *Principia*, written out and sent on to Halley after the original MS had been completed and forwarded to London. There is no extant early MS of Newton's containing this derivation, however, and we have no evidence as to the date when Newton found this particular derivation, which had to be later than the mid-1660's and earlier than 1686.

12 The existence of the relevant documents was announced in Turnbull (1953), and the documents were edited, translated, and interpreted in Newton's *Correspondence*, of which Turnbull was editor, Vol. 1, pp. 297–303, Vol. 3, pp. 46–54. These same documents are printed and discussed in Hall (1957), Herivel (1961), and Herivel (1965).

13 In this computation, Newton first computed the rectilinear distance through which a body would be accelerated uniformly by the constant action of a force equal in magnitude to the centrifugal endeavour at the earth's surface, during the time of one complete revolution of the earth (24 hours). This proves to be 19.75 × earth's radius. Using Galileo's law, and the value of 3497 miles for the earth's radius, the distance fallen in one second turns out to be 5/108 ft, which is 1/350th of 16 ft, the average distance of free fall on the earth's surface in one second. Hence Newton's conclusion, that the magnitude of centrifugal endeavour at the earth's surface is only 1/350th as great as gravity, depends on the value taken for the earth's radius. A more correct value (4000 miles, rather than 3497 miles) would have given the result of 1/286 rather than 1/350. For the significance of this difference, see note 14 *infra.*

14 Using the correct value of 1/286, the result would be that gravity is 3575 times as great (12 1/2 × 286 = 3575) as the centrifugal endeavour of the moon. This is very close to 3600 (off by 0.7%), whereas 4400 differs significantly from 3600 (the difference is 800 or 22%).

15 Newton's *Correspondence*, Vol. 1, p. 290; Newton had just received a presentation copy of Huygens's *Horologium oscillatorium*, containing a statement of the rule for centrifugal force.

16 See note 18 *infra.*

17 Even if it had occurred to Newton that the inverse-square rule for planetary endeavour of receding might apply to the moon, the discrepancy between 4400 (or '4000 or more') would have been so discouraging that he might not even have bothered to write it down. But this is all conjectural and hypothetical. As D. T. Whiteside has stressed (in his edition of Newton's *Math. papers*, vol. 6, p. 5, n. 13), 'The first documented record of Newton's making such a test [as described in the later autobiographical memorandum] of the inverse-square variation of terrestrial gravitation with distance is an off-hand reference to a recent "computation" of his to that end which he set down about January 1685, without citing any details, in the augmented scholium to Problem 5 of his revised preliminary tract 'De motu Corporum' . . . '

The autobiographical draft memorandum was written for Des Maizeaux in about 1718 about half a century after the supposed moon test of the 1660's. Newton also recounted these events to Whiston in the early 1690's and again to Pemberton in the 1720's. Thus even Whiston's version dates from more than a quarter of a century after the alleged events. To these accounts we may apply the inverse-square law of the Newtonian John Craig, that their credibility decreases as the square of the time since the events are supposed to have occurred.

[18] It may also be observed that in the later Des Maizeaux autobiographical memorandum, Newton says that he had combined with Kepler's third law the rule for estimating 'the force with which a globe presses the surface of a sphere'. But in the document of the late 1660's, Newton does not use this approach (which occurs in the 'Waste Book', see note 10 *supra*), but rather deals with the distance through which a body travels out along the tangent as a result of its centrifugal force. This further discrepancy may suggest either (a) that the Des Maizeaux memorandum is wrong on yet another count, or (b) that there is (or was) another document in which Newton proceeded in the manner described. Could this have been the document containing the derivation of the centrifugal force law for a body pressing a circle within which it travels, and from which Newton drew the final paragraph for the scholium to prop. 4, bk. 1, as described in his letter to Halley in 1686 (see note 10 *supra*)? Who knows? No such document exists among Newton's papers. It is folly to try to guess about the contents of a document for which we have no hard evidence as to its existence. If there were such a document, and it gave a better result than the one from the late 1660's, it would no doubt be dated in the early 1670's, for which time we have no other record of the kind of thinking and research from which such a result would have followed. In the present context, in any event, we would guess that any such document would have been still couched in terms of an endeavour of receding or a centrifugal force and not yet the later centripetal force.

[19] Whiston's *Memoirs* (London, 1749), vol. 1, pp. 35–38, reprinted in Ball (1893), pp. 8–9.

[20] Pemberton (1782), preface.

[21] As ref. 19.

[22] As ref. 20.

[23] In describing these events in a letter to Halley, 20 June 1686, Newton referred to 'one of my papers' in which 'the proportion of our gravity to the Moon's *conatus recedendi a centro Terrae* is calculated though not accurately enough'; *Correspondence*, vol. 2, p. 436.

[24] See notes 13, 14 *supra.*

[25] As ref. 19.

[26] Whiteside (1970), pp. 11–12, Whiteside (1976), p. 318; see also §2, n. 5, *supra*.

It should be observed that what Whiteside discovered about the lunar vortex confirms only that part of Whiston's statement about '*Cartesius's* Vortices' and not the rest of his report about the alleged moon test. It is not inconceivable that Newton may have thought about the possibility that the inverse-square law that he had derived for the centrifugal conatus of planets might also hold for the moon; and that he would then have realized at once that 3600 was far from 4400 and so would not even have bothered to write down his conjecture and its obvious failure to fit the facts. Following Whiteside's disclosures, however, we would have to stress that in the 1660's Newton believed firmly in the real existence of a kind of centrifugal conatus or force that in the non-circular case was not even quantitatively equal to an inverse-square central force. Hence, Whiteside's analysis leads to the conclusion that a moon test which would have come out exactly in the 1660's would at best have only made Newton wonder why for a lunar elliptical orbit the centrifugal force in that orbit should be approximately equal to the central inverse-square force that holds for circular orbits. See Whiteside (1964), esp. n. 54.

[27] See, further, n. 6 to §7 *infra.*

§4: PRELIMINARIES TO UNIVERSAL GRAVITY, 1679–1684

[1] Newton, however, continued to use the old expression of "conatus", even in the *Principia*. For example, in the scholium on space and time at the end of the Definitions, he uses the word "conatus" in relation to the "endeavour of receding" of rotating bodies. He says of a rotating pair of globes connected by a cord: 'innotesceret ex tensione fili conatus globorum recedendi ab axe motus' ('the globes' endeavour of receding from the axis of motion would be known from the tension of the cord'). In the scholium following prop. 53, Bk. 2, he talks of the 'conatus recedendi a centro' of fluid particles. By this time he understood that such an apparent "endeavour" of receding is merely a manifestation of the inertia of a body, which tends to carry the body along the tangent, effectively outward from the center. For the scholar the presence of such archaic expressions, used in a new sense, is a sign of the sources of ideas that were of importance in the development of the scientific thought of Newton: archaeological reminders of origins of concepts.

On the importance of the shift from centrifugal to centripetal force, see n. 6 to §7 *infra* and also n. 29 to §6.

[2] See Koyré (1952), Lohne (1960), Westfall (1967*a*), Westfall (1971), Whiteside (1964*a*), Whiteside (1970), Herivel (1965), and the autobiographical documents of Newton's printed in Cohen (1971), suppl. 1. See also Patterson (1949, 1950). See, especially, Cohen (1980), §5.4, and Cohen (1981).

[3] See §7 *infra*. As to how Newton could continue to think simultaneously of aethereal vortices and centripetal forces, see §6 *infra.*

[4] Newton's letters to Halley in *Correspondence*, vol. 2, display the various kinds of evidence Newton could adduce for having known an inverse-square law (but, once again, let me insist, not an inverse-square law of *centripetal* force) before Hooke's letter.

[5] They had earlier had a controversy in relation to Newton's theory of light and colors.

[6] Hooke to Newton, 24 Nov. 1679, *Correspondence*, vol. 2, pp. 297–298.

[7] Newton to Hooke, 28 Nov. 1679, *Correspondence*, vol. 2, p. 300.

[8] *Ibid*. See Lohne (1967).

[9] Hooke to Newton, 9 Dec. 1679, *Correspondence*, vol. 2, p. 305. Toward the conclusion of the letter, Hooke said: 'I could adde many other conciderations which are consonant to my Theory of Circular [i.e., circulating or orbiting] motions compounded by a Direct motion and an attractive one to a Center'. See Koyré (1952).

[10] 13 Dec. 1679, *Correspondence*, vol. 2, p. 307.

[11] Whiteside (1964*a*), p. 134. Newton did not rise to the bait cast in the sentence quoted in note 9 *supra.*

[12] Hooke to Newton, 6 Jan. 1679/80, *Correspondence*, vol. 2, p. 309, second sentence of letter.

[13] Hooke to Newton, 17 Jan. 1679/80, *Correspondence*, vol. 2, p. 313.

14 *Ibid.*

15 They are gathered together in Cohen (1971), suppl. 1.

16 There are two reasons why we may have more than ordinary confidence in the truth of these memoranda with respect to Hooke. First of all, for Newton to have given credit to Hooke (even by implication), despite the bitterness of the controversies between them, has more weight than if Newton had been writing of someone with whom he had not been at odds. Second, he was not using this event to establish a priority with respect to dynamics, but rather was trying to set forth a chronology of ideas. In actual fact, he was saying that he had not really understood orbital motion until he was almost forty years old.

17 Cohen (1971), p. 293, quoted from U.L.C. MS Add. 3968, f. 101. Although Newton mentions that the occasion for his demonstration was the correspondence with Hooke, he does not specify that Hooke's genuine contribution was the proposal that orbital motion be conceived as compounded of an inertial component plus a constant acceleration (or force) directed to a focus. This particular contribution of Hooke's may be seen in his letters to Newton.

Newton also developed and used an ingenious quantitative measure of a centripetal force; see notes 36, 37, and 40 *infra*. It is to be observed once again that what Newton had found earlier was not the law of the inverse square for 'the forces which kept them [i.e., the planets] in their Orbs about the Sun', but rather the planets' 'endeavours of receding from the sun', which only later became 'the forces which kept them in their Orbs'.

18 In the *Principia*, as in the prior *De motu*, Newton does not have a proposition stating that if there is an inverse-square force the orbit will be one of the conic sections. In the *Principia*, Newton proves that if the orbit is an ellipse (prop. 11, bk 1), a hyperbola (prop. 12), or a parabola (prop. 13), then the force will vary inversely as the square of the distance. The converse is the subject of corol. 1 to prop. 13, where – in the first edition – Newton merely states (without proof) that if the force varies inversely as the square of the distance, the orbit will be a conic section having a focus at the center of force. In the second and third editions, he introduces the outline of his proof. For details, see Whiteside's notes on pp. 147–149 of vol. 6 of his edition of Newton's *Math. papers*, supplemented by his notes on pp. 348–350; and also Aiton (1964). Prof. Robert Weinstock has recently challenged the validity of this proof; see, as a first essay, J. C. Rainwater and R. Weinstock, 'Inverse-square orbits: a geometric approach', *American Journal of Physics*, 1979, vol. 47, pp. 223–227, esp. n. 6. This question is discussed also in a most important article by R. C. Riddell, 'Parameter disposition in planetary theories and its relation to Newton's *Principia*', *Archive for History of Exact Sciences*, 1980, vol. 23, pp. 87–157, esp. pp. 155–157.

Newton may have believed that the converse of what he had proved in props. 11–12–13 was so obvious as not to require proof. On this score it will be of real interest to see Prof. Weinstock's fully developed argument. I would find it difficult, however, to find Newton guilty of believing that if conic orbits imply an inverse-square force then it must follow – without any need for proof – that inverse-square forces imply conic orbits. The absence of a proof in corol. 1 to prop. 13 in the first edition of the *Principia* would not mean (for me) that Newton did not believe a proof was needed, nor that he would not construct a proof. Rather, he may have thought the proof was obvious and needed no explicit statement, such as he added in the later editions. It is a separate question, of course, as to whether that proof is valid.

Yet it must be admitted that in Newton's day, every Newtonian did not make a sharp and clear distinction between the statement concerning the orbit produced by an inverse-square force and the force required to produce an orbit that is a conic section. Thus, in John Conduitt's memorandum of Halley's visit to Newton, the one that began the train of events leading to *De motu* and the *Principia*, it is said that Halley asked Newton, 'what he thought the curve would be that would be described by the planets supposing the force of attraction towards the sun to be reciprocal to the square of their distance from it' – to which 'Sir Isaac replied immediately that it would be an ellipsis'. When asked 'how he knew it', Newton replied, 'I have calculated it'. The subsequent write-up was presumably the tract *De motu*, in which Newton proves the converse of what Conduitt records, namely, he finds the force corresponding to an orbit that is a conic. For Conduitt's text, see Cohen (1971), pp. 297–298.

19 In a letter to Halley, 14 July 1686, Newton said that Hooke's 'Letters occasioned my finding the method of determining Figures, which when I tried in the Ellipses, I threw the calculation by, being upon other studies & so it rested for about 5 years till upon your request I sought for that paper, & not finding it did it again & reduced it into the Propositions shown you by Mr Paget'. That is, when Halley visited Newton in August 1684 and asked him what the orbital curve would be if planets were attracted by an inverse-square force, Newton replied that it would be an ellipse, on the basis of calculations made five years earlier, in 1679 (which in the old style of reckoning extended into the first three months of 1680). Not being able to find those calculations, he rewrote them in the form of the tract *De motu*. On the date of Halley's visit, see Cohen (1971), pp. 54–56. What Newton proved, however, in *De motu* was not that if the attractive force varies as the inverse square of the distance, then the orbit will be an ellipse; rather, he proved the converse, namely, that if the orbit is an ellipse, then the attractive force will vary inversely as the square of the distance. See, further, note 18 *supra*.

20 These revisions are discussed in § 5 *infra*.

A copy of this tract was taken to London by Edward Paget in November 1684 and was apparently the one copied into the Register Book of the Royal Society in December 1684; it may still be found in Register Book, vol. 6, pp. 218–234. This draft was published in Rigaud (1838). For details, see Cohen (1971), p. 56, and Whiteside's note 2 on pp. 30–31 of vol. 6 of *Math. papers*. See also Ball (1893) and Hall & Hall (1962) for printings of this tract and discussions of its significance. A fourth printing, with extensive commentary, occurs in Herivel (1965). The various versions are also printed with commentary in *Math. papers*, vol. 6, pp. 30–91.

21 There are two MS versions of this short essay. One is Newton's original holograph MS (U.L.C. MS Add. 3965, fols. 1–4), first published (incompletely) in Ball (1893), pp. 116–120, and published in full in Hall & Hall (1962). The other version was published in King (1829).

22 On this topic, see Hall & Hall (1962), pp. 293–301, Herivel (1969*b*), Hall (1963), Herivel (1965), pp. 247–256, Westfall (1969), Westfall (1971), p. 430. For the text, see Hall & Hall (1962), pp. 293–301; *Correspondence*, vol. 3, pp. 71–77. In this tract, prop. 1 displays the law of areas as a consequence of a central force much as in *De motu* and the *Principia*. The chief difference lies in prop. 2, showing that elliptical orbits imply an inverse-square force. Westfall (1969) summarizes Herivel's argument (1965; pp. 108–117), first advanced by Herivel in 1961, 'that a certain paper among the Portsmouth collection was Newton's original demonstration that a body revolving in an elliptical

orbit . . . must be attracted by a force that varies inversely as the square of the body's distance from the attracting center'. After noting that 'no one . . . has heretofore accepted Herivel's identification', Westfall (p. 52) stated unequivocally, 'I believe I rank as his first convert'.

23 There is, however, neither a letter from Newton to Locke mentioning this essay, nor a letter from Locke to Newton acknowledging the receipt of the essay. Whiston published a Latin version of the essay in 1710, reprinted in 1726, and an English version in Whiston (1716), pp. 159–165.

24 The watermark on the paper on which Newton's holograph version is written occurs in the mid-1690's; even if such dating may be off by as much as ten years, the paper would assign a date to this version of about the middle 1680's; it would not then have been composed in 1679. That is, if this version had been Newton's primitive or first essay, composed at the time of the correspondence with Hooke, then he would have had to copy it out later in his own hand. In any event, an amanuensis made a copy, which 'differs somewhat' from this version, for transmission to John Locke. That copy, endorsed 'Mr Newton, Mar 89/90', was found among Locke's papers and published in 1830 in the second edition of Lord King's *Life of Locke* (1829), pp. 209–14, (1830), vol. 1, pp. 389–400, (1972), pp. 210–216, and is reprinted in Newton's *Correspondence*, vol. 3, pp. 71–77.

In the context of the present discussion, it is not of any real import as to whether the essay was written before or after *De motu*. What is of significance is primarily the state of Newton's understanding of celestial dynamics in 1684 and 1685, when he stood on the brink of the discovery of universal gravity – a concept that he had not then fully formulated.

25 D. T. Whiteside has pointed out that this proof 'is more subtle and considerably less cogent than it may at first appear'; see his note 19 on pp. 35–37 of Vol. 6 of his edition of Newton's *Math. papers*. In that extensive note, Whiteside lays bare the infinitesimal assumptions, including that the orbital arc 'be itself infinitesimally small'.

26 See Maeyama (1971), Whiteside (1964*a*). This method of approximation is set out in Streete (1661), one of the sources where Newton would have encountered it and learned how to use it, and where he also encountered Kepler's third law. It is significant that Streete discusses the law of elliptical orbits and the harmonic law, but does not even mention the law of areas. For a discussion of the approximation used in place of the law of areas, see Cohen (1980), §5.2.

27 On the fact that there were some astronomy books and manuals presenting the area law, see Russell (1964).

D. T. Whiteside (*Nature*, 1974, vol. 248, p. 634) states that 'after my own nearly twenty years of continuous study of Newton's scientific papers (including many in private possession which are not generally accessible) I will reiterate that I have as yet found nothing which establishes in any way that Newton was, before 1676, aware of the law's verbal enunciation or begins to hint that he "deeply pondered" its meaning at any time before late 1679'. This is a restatement of the position first set forth in Whiteside (1964*a*). The point at issue is not, of course, whether Newton had ever encountered or had ever read the law, as he may well have done in looking over his 'well-thumbed' copy of Nicolaus Mercator's *Institutiones astronomicae* of 1676, of which he had a copy in his personal library, or whether he may have become aware of this law from encountering a passing reference to it in Christopher Wren's tract on Kepler's problem, which was published as a supplement to John Wallis's *De cycloide* of 1659, which Newton had certainly studied; for which see Cohen (1980), pp. 342–343. Indeed, it is not simply a question of Newton's 'presumed ignorance of the law's statement' up to 1679, but rather 'his recognition of its theoretical necessity in his proof of the elliptical orbits of the planets under the condition (as Hooke laid it down in their correspondence in the early winter of 1679/80) that their undeviated motion be uniform and in a straight line'.

I have expressed my own opinion on this matter in a somewhat different way in *Nature*, 1974, vol. 250, p. 180, to the effect that there is no documentary evidence that Newton was consciously aware of Kepler's law of areas as a physical principle before 1679. This law appears explicitly in Newton's writings for the first time in *De motu* in 1684, although Newton may have made use of it after (or during) his correspondence with Hooke in 1679–1680, when (he later said) he solved the problem of elliptical motion. See, further, Cohen (1980), p. 228 and notes 10–12 on pp. 342–343.

[28] Mercator (1676), pp. 286–287. Mercator (p. 286) wrote of Newton as a man 'cujus humanitati hoc & aliis nominibus plurimum debere me lubens profiteor'. Newton's letter to Mercator is lost, and its contents are known only through Mercator's printed summary. On Newton's theory of the libration of the moon, see *Principia*, Bk 3, prop. 17.

[29] In Mercator's hypothesis, the turning radius vector for equal angular motion 'is centered in the line of apsides at a distance from the sun roughly 5/8 times the doubled eccentricity', a ratio set by Mercator 'exactly equal to the "divine section" $(\sqrt{5} - 1)/2$'; cf. D. T. Whiteside on Mercator in *Dictionary of Scientific Biography*, vol. 9, p. 310.

[30] Wren's essay on this subject ('De recta tangente cycloidem primariam') was published as an appendix to John Wallis's tract (1659), *De cycloide*, pp. 70–74 [= 82]; the concluding section, pp. 80sqq., is entitled 'De problemate Kepleriano per cycloidem solvendo'.

Wren writes that from Kepler's rule that the orbital speed of a planet is inversely proportional to its distance from the sun, 'he [Kepler] ingeniously contrived the following hypothesis. That is, by lines drawn [from the ellipse] to the sun, he cuts the area of a planetary ellipse into equal mixtilineal triangles, whence it results that the curve of the ellipse is divided into unequal portions [or arcs], smaller ones in the neighborhood of aphelion and larger ones in the neighborhood of perihelion; and he supposes that the planet is carried through these [unequal arcs or] portions in equal times. Wherefore, in order to obtain the co-equated anomaly from the mean motion, a semi-ellipse has to be cut through a focus in a given ratio; or (which he demonstrates to be the same thing) a semicircle is to be cut through any point of its diameter in a given ratio. It is remarkable how much Kepler has sweated over this problem, "revolving" his orbits "with great exertion and he gets nowhere at all". At length, puffing and panting, he tearfully implores the aid of geometers, fearing meanwhile lest the problem be found unsolvable [*inexplicabile*] because of the ἑτερογένεια [heterogeneity] of arcs and sines. Which problem we ourselves, nevertheless, demonstrated long ago with the aid of a protracted cycloid, as follows'. The quotation given by Wren is from a bit of poetry based on a passage in Cicero's *Tusculan disputations*. Wren's solution of Kepler's problem was essentially reproduced by Newton in the scholium following prop. 31, bk 1, of the *Principia*. See, further, Cohen (1980), pp. 228, 342–343.

[31] As has been mentioned, there is no primary documentary evidence as to when the sequence of propositions in *De motu* and in the *Principia* was first established. The first dated document in which this sequence appears is the version of *De motu* entered in the

Royal Society's Register Book under the date of November 1684. But in various MSS Newton says that this order of propositions corresponds to the solution of the problem of elliptical orbital motion which he achieved following (and as a result of the stimulation of) the suggestions made by Hooke during their correspondence in 1679/80.

32 In the *Principia*, the intervening props. 2 & 3 deal with the converse of prop. 1 (that is, with the area law leading to a centripetal force) and the situation in which the orbiting body and its center may be accelerated.

33 U.L.C. MS Add. 3968, § 27, fol. 415: 'Mr Huygens gave the name of vis centrifuga to the force by which revolving bodies recede from the center of their motion. Mr Newton in honour of that author retained the name & called the contrary force vis centripeta'; printed in Koyré & Cohen (1962), pp. 122–123. That is, Newton invented the name "vis centripeta" so that it could be paired with "vis centrifuga", and he wanted to keep the name "vis centrifuga" (although he now explained this centrifugal tendency in terms of inertia) so as to honor Huygens.

34 Newton to Leibniz, 16 Oct. 1693, *Correspondence*, vol. 3, p. 285, par. 3: 'Quae vir summus Hugenius in mea notavit ingeniosa sunt'.

35 Only the earth and Venus have orbits that can be considered approximately circular; all other planets (and the moon) have such obviously elliptical orbits that this condition of circularity of prop. 4 (theor. 2) cannot really apply to them. Of course, the proof that the centripetal force in circular orbits varies inversely as the square of the distance also depends on the acceptance of the concept of centripetal (as opposed to centrifugal) force.

36 In the *Principia*, the intervening prop. 5 discloses a means of finding the center of forces when the velocities are given at several points in the orbit.

37 See Whiteside (1970*a*), pp. 122–123. Possibly this concept of the measure of a force (prop. 6 of the *Principia*) is to be dated in 1684 rather than 1679–80; see note 40 *infra*.

38 He then shows separately that parabolic and hyperbolic orbits also require an inverse-square force.

39 In the *Principia*, the intervening propositions discuss (prop. 7) the centripetal force directed to any point in a circular orbit (equivalent to problem 1 of *De motu*), then (prop. 8) motion in a semicircle when the center of force is very far away, then (prop. 9) motion in a spiral, then (prop. 10) motion in an elliptical orbit with the force tending toward the center (problem 2 of *De motu*).

40 In a MS memorandum (U.L.C. MS Add. 3968, fol. 106), printed in Cohen (1971), p. 295, Newton said: '... the 1st & 11th [Propositions were composed] in December 1679, the 6th 7th 8th 9th 10th 12th 13th 17th Lib. I & the 1 2 3 & 4 Lib. II in June & July 1684'.

41 This is prop. 15, Bk 1, of the *Principia*. Since it does not occur in Newton's list of propositions in the MS quoted in note 40 *supra*, possibly it was not part of the Newtonian extension of the physics of circular orbits to elliptical orbits until he actually wrote up his results to send to the Royal Society in November 1684.

42 The title reads: *Astronomia nova αἰτιολόγητος seu physica coelestis* ... [*A new astronomy based on causes, or a celestial physics* ...]

43 Hooke to Newton, 6 Jan. 1679/80, *Correspondence*, vol. 2, p. 309.

44 Aiton (1969*a*); see also Wilson (1970).

45 At the apsides, and only at the apsides, the tangent is perpendicular to the radius vector from the focus to the planet's position so that the two distances are there equal.

[46] As the first corollary to the first proposition of the first Book, this statement of the true speed law was placed in the most prominent position possible. Of course, we do not have any documentary evidence that it was Newton's continued animus against Hooke that caused him to place this statement in such a primary position.

§5: THE ESSENTIAL STEP TO UNIVERSAL GRAVITY IN 1685

[1] See note 20 to §4 *supra*.
[2] Scholium following prob. 3, trans. in Hall & Hall (1962), p. 277; also, Herivel (1965), p. 282 and Whiteside (ed.), *Math. Papers*, vol. 6, p. 49.
[3] See Duhem (1954), pt. 2, ch. 6, sect. 4, pp. 190–193; also Popper (1972), pp. 200–201; Cohen (1974).
[4] The problem here is one of chronology without identifiable documents. After Halley's visit to Newton in August 1684, Newton sent a tract to London, using Paget as messenger. This tract was described by Newton in a letter to Halley, 14 July 1686 (*Correspondence*, vol. 2, pp. 444–445) as the calculations which 'I ... did ... again and reduced ... into the Propositions shewed you by Mr Paget'. Halley then went to Cambridge to see Newton for a second time, and reported to the Royal Society that he 'had lately seen Mr. Newton at Cambridge, who had shewed him a curious treatise, *De Motu*'; at 'Halley's desire', Newton 'promised ... [it would] be sent to the Society to be entered upon their Register'. Concerning this, see Rigaud (1838), p. 15; Ball (1893), p. 30; Cohen (1971), p. 55. I presume that the tract sent via Paget in November was the same as that sent in December and copied into the Register Book. Halley's statement does not indicate whether the text that Newton showed him was a revised version of that first draft; but Halley does not say that he saw again the 'curious treatise, *De Motu*', so possibly he saw a revised and not quite finished version.
[5] A one-body "system" is, of course, strictly impossible. But I use the word here so as to maintain a continuity of expression with "two-body system", "three-body system", and "many-body system". What I am referring to here can, however, be considered as a system in the sense that it consists not merely of an isolated body, but of such a body plus a center of force. See, further, note 8 *infra* and also §7 *infra*.
[6] The law of action and reaction is a generalization and transformation of a principle that had been developed in relation to collisions or impacts. Newton's originality – like the originality of all innovations, discoveries, and revolutions in science – consisted of a degree of transformation of existing scientific ideas; in this case to make a transformation of a rule for impacts or collisions (that is, for contact forces) so that it became a general law for any kinds of forces whatever: impacts, pressures, and even centripetal forces of "attraction". On this topic, see Cohen (1980). Concerning the originality of the third law, see Mach (1960), p. 243, ch. 2, pt. 3, §11.
[7] That is, all too many writers have not recognized that the third law posits that the "action" of a body A on a second body B is equal and opposite to the "reaction", which is the action of B on A. Such a situation cannot produce an equilibrium; equilibrium results when equal and opposite forces act simultaneously on a single body (A or B).
[8] It could be said that Newton had reduced the solar system to a set of separate or individual systems, one planetary mass point in each, with a common center of force in the sun. They are separate to the extent that the motion of each such planetary mass

point is not affected by the others; there are no inter-planetary forces of perturbation. These are essentially mass points because their behavior is wholly independent of such physical characteristics as size or shape.

9 Intro. to *Principia*, sect. 11, Bk 1.

10 The dating is precise, since the mutual attraction of the two bodies occurs in an addition to *De motu* which is not present in the Royal Society version (Dec. 1684). Before long, in 1685, he was deep in the composition of the *Principia*, and *De motu* was far behind.

11 Of course, this step implies that all planets are essentially the same, at least with respect to gravitation, and that similarly there is no difference between the sun and planets, even though the sun shines by its own light and is extremely hot. Newton actually made this condition precise when he found that the only property that counts in considerations of gravitation is mass. In Book 3 of the *Principia* (prop. 5, corol. 1), he said: 'No one doubts that Venus, Mercury, and the rest are bodies of the same kind as Jupiter and Saturn.'

In the revised *De motu*, Newton recognizes that the power of attracting and being attracted holds for the case of two planets just as much as for a planet and the sun. By extension, as he was to find next, this same mutual attraction may also occur in the case of a planet and its satellites, and hence a planetary satellite will both attract and be attracted by its own planet, by the sun, and by other planets (and even by their satellites). The argument will eventually lead Newton (in the *Principia*) to any kind of terrestrial object attracting and being attracted by the earth and so attracting and being attracted by other such terrestrial objects: and so on to universal gravity. But in *De motu*, Newton does not get beyond interplanetary attractions.

12 Addition to *De motu* in the revised and expanded version; Hall & Hall (1962), p. 281 and Herivel (1965), p. 301.

13 *Ibid.*

14 See, further, §6 *infra.*

15 See Cohen (1981) and, for a more extended discussion, Cohen (1980), §5.6.

§6: THE "NEWTONIAN STYLE"

1 This thesis is developed in Cohen (1980).

2 Husserl (1970), pt. 2, §9.

3 This accords with the general theory of scientific advance, and even of scientific revolution, developed in Cohen (1980) and Cohen (1982), according to which even revolutionary innovations turn out to be transformations or extensions of existing scientific ideas and methods.

4 See note 5 to §5 *supra.*

5 See Whiteside (1966*a*), Whiteside (1970), and Whiteside (1970*a*).

6 Prop. 66, Bk 1.

7 Prop. 69, Bk 1.

8 Sect. 12 & 13, Bk 1.

9 Bk 2.

10 Sect. 10, Bk 1.

11 Book 3 is called 'De mundi systemate. Liber tertius.'

12 Introduction to *Principia*, Book 3.

13 On the change from 'hypotheses' to 'phaenomena' and 'regulae philosophandi', see Koyré (1955), Koyré (1960), Cohen (1953), Cohen (1956), pp. 130–134, and Cohen (1966).

14 Scholium following prop. 52, Bk 2.

15 He considered bodies moving both in fixed orbits and in orbits that were themselves in motion (sect. 9, Bk 1).

16 Prop. 23, Bk 2. He also generalized the law of inter-particle force which led to a consequent generalized form of Boyle's law.

17 Sect. 14, Bk 1.

18 Sect. 8, Bk 2.

19 Prop. 30, Bk 2; sect. 10, Bk 1.

20 Sect. 10, Bk 1; sect. 6, Bk 2.

21 Sect. 1, 2, 3, 4, 7, 9, Bk 2.

22 *Correspondence*, vol. 2, p. 440. In this letter Newton told Halley that he had used this same hypothesis of 'uniform gravity' in 1679, when writing to Hooke about 'the descent of a body down to the center' of the earth. At that time, he said, he knew that outward 'from the superficies of the Planet', gravity must vary inversely as the square of the distance from the center; and he also knew full well that this inverse-square law does not hold within the earth. Hence, in considering motion downward from the earth's surface, 'I used not the duplicate proportion'.

In a letter to Halley, 27 July 1686 (*Correspondence*, vol. 2, p. 447), Newton said further of a hypothesis he had used in a mathematical computation, 'Whether this be true is not material'.

23 These are the only two gravitational forces occurring in nature. The direct distance law applies, as Newton proved, to the force exerted by a homogeneous sphere on a point at a distance r from the center. The outer spherical shell was proved to exert a null force on any interior particle. The inner sphere of radius r acts as if all of its mass were concentrated at the center. Hence the force is directly proportional to the mass of the inner sphere (which, being homogeneous, is proportional to r^3) and inversely proportional to $1/r^2$; hence this force is proportional to r. See *Principia*, props. 70–78. Bk 1, and props. 8–9, Bk 3.

24 On the significance of prop. 41, see Aiton (1964), and Whiteside's commentary in *Math. papers*, vol. 6, pp. 344–356, esp. pp. 349–350 (n. 208).

25 We may see the expression of the Newtonian style also in the concluding scholium to this sect. 11, Bk 1. Here Newton says again that he is using the word "attraction" in 'a general sense for any endeavor whatever of bodies to approach one another', whatever be the mode by which that may occur, whether 'by the action of the bodies either mutually seeking one another or agitating one another by means of spirits emitted' or by 'the action of aether or of air or of any medium whatsoever – whether corporeal or incorporeal – in any way impelling toward one another the bodies floating therein'. And he explains that the word "impulse" is used by him 'in the same general sense', since he is considering here 'not the species of forces and their physical qualities but their quantities and mathematical proportions'.

26 See Koyré (1965), pp. 57–58, and also pp. 151–153.

27 On Newton's restricted use of the word "attraction", see Cohen (1980), pp. 82–83.

28 At the end of the discussion of centripetal forces, following Def. 9, Newton explains:

'I use the words *attraction, impulse*, or *propensity* of any sort whatever toward a center for one another without distinction and indiscriminately, considering such forces not physically but merely mathematically. Hence let the reader beware of thinking that by words of this kind I am anywhere defining a species or mode of action or a physical cause or reason; or that I am attributing forces truly and physically to centers (which are mathematical points), if I happen to say that centers attract or that centers have forces.'

29 In the *Discours sur la cause de la pesanteur* (1691), written by Huygens in response to the *Principia*, Huygens explained that he himself 'had not extended the action of gravity to such great distances as those between the sun and the planets, or between the moon and the earth' because 'the vortices of M. Descartes which formerly seemed to me rather likely, and which I still had in mind, got in the way'. Here is an example of the way in which the lack of a "Newtonian style" could prevent a sagacious mind from achieving its maximum potential creativity. See Koyré (1965), p. 122; Huygens's *Oeuvres*, vol. 21, p. 472.

30 See note 15 to §4 *supra.*

31 See Guerlac's article on Hauksbee in the *Dictionary of Scientific Biography.*

32 *Opticks*, Qu. 21 (first published in second English edition of 1717/18).

§7: NEWTON'S VIEWS ON THE AETHER AT THE TIME OF WRITING THE *PRINCIPIA*

1 This experiment was described by Newton in the essay, 'De aere et aethere', Hall & Hall (1962), pp. 227–228; it was referred to in his hypothesis of 1675, 'explaining the Properties of Light', Cohen & Schofield (1958), pp. 179–180, Birch (1757), vol. 3, pp. 249–250. For details, see Westfall (1971), pp. 336, 374, and Westfall (1970).

2 *Principia*, book 2, conclusion of General Scholium at the end of sect. 7 (in the second and third editions; at the end of sect. 6 in the first edition).

3 There is no direct or documentary evidence as to when this experiment had been performed; see Cohen (1971), pp. 101–104. I would set the date of this experiment in 1685, since the theoretical underpinning of the experiment and Newton's interpretation of his results are based on a rather advanced and difficult train of thought concerning the dynamics of pendulum motion, of which we have no trace whatever in Newton's papers earlier than the composition of Book 2 of the *Principia* (see note 26 *infra*).

A discussion of the significance of this experiment is to be found in Westfall (1970). The experiment did not prove that the aether does not exist, but only that the resistance of an aether (if any) is too little to manifest itself in pendulum experiments. But it was interpreted by Whiston (in 1725) as proving that '. . . 'tis certain there is no *Subtile Matter* pervading the Universe, as some have supposed'; the reasons are that 'pendulous Bodies receive no sensible Resistance through the Aethereal Regions' and that 'Comets move prodigiously swift, with the utmost Freedom, and without any sensible Resistance through the Aethereal Regions'. Quoted in Westfall (1971), p. 410.

4 On Newton's early concept of an aethereal "shower" that could cause gravity (i.e., heaviness toward the earth), see Westfall (1971), pp. 329–331.

5 The suggestion that an aether of varying density might be the cause of terrestrial gravity occurs in the penultimate paragraph of his letter to Boyle, 28 Feb. 1678/9, reprinted in Cohen & Schofield (1958), p. 253. It should be noted that at this time Newton was only concerned about bodies being "heavy" toward the earth.

[6] In writing about Newton's *Principia*, Huygens said that he could 'not agree with a Principle according to which all the small parts that we can imagine in two or several different bodies mutually attract each other or tend to approach each other' because 'the cause of such an attraction is not explainable by any of the principles of Mechanics, or of the rules of motion'. Even more important, Huygens was not 'convinced of the necessity of the mutual attraction of whole bodies, since I have shown that, even if there were no Earth, bodies would not cease to tend toward a center by that which we call gravity'. Quoted from the translation given in Koyré (1965), p. 118. The original comes from Huygens's *Discours sur la cause de la pesanteur* (Leiden, 1960), p. 159; *Oeuvres complètes*, vol. 21, p. 471. See, also, n. 29 to §6 *supra*.

It is to be observed, furthermore, that Huygens's argument shows how crucially important it was for the development of the concept of universal gravity that Hooke gave Newton the clue to a centrally-directed planetary force, to replace the older centrifugal force. A planetary (or any other) centrifugal force concentrates attention on the orbiting body, whereas a centripetal force draws attention to both the orbiting body and the central body toward which the force is directed. Motion under the action of a centrifugal force thus – like motion in vortices – would not have led to a mutually interacting force and thus is in direct contrast to the action of a centripetal force.

[7] See Whiteside (1964*a*), p. 127, and Whiteside (1976), p. 318. In discussing the aether in his hypothesis of 1675, Newton wrote of 'the aethers in the vortices of the sun and planets'. This hypothesis is printed in Birch (1757), vol. 3, pp. 248–260; and in Cohen & Schofield (1958), pp. 178–190, 250–254.

[8] In Westfall (1970), Westfall (1971); see the various articles by Guerlac in the bibliography, now collected in book form in Guerlac (1977).

[9] See Guerlac's article on Hauksbee in the *Dictionary of Scientific Biography*. On electricity in relation to the final paragraph of the concluding General Scholium to the *Principia*, see Hall & Hall (1959), Koyré & Cohen (1960).

[10] Apart from references to a medium, such as the 'medium, if there should be any, freely pervading the interstices between the parts of bodies' (in def. 1), there are a number of direct references to aether in the *Principia*. Toward the end of the scholium to the Laws of Motion, Newton writes (without raising any question of existence) about 'the whole earth, floating in the free aether'. In corol. 2, prop. 6, Bk 3, he writes: '. . . if the aether or any other body whatever either were entirely devoid of gravity or gravitated less in proportion to the quantity of its matter . . .' Toward the conclusion of the scholium at the end of sect. 8, Bk 2, he concludes that sound consists in 'the agitation of the whole air', and not 'in the motion of the aether or of a certain more subtle air'. At the end of the scholium with which sect. 11, Bk 1, concludes, he says: 'I use the word *attraction* here in a general sense for any endeavour whatever of bodies to approach one another, whether that endeavour . . . arises from the action of aether or of air or of any medium whatsoever – whether corporeal or incorporeal – in any way impelling toward one another the bodies floating therein'. In discussing the tails of comets in lemma 4, Bk 3, he states his opinion that they 'arise either from the reflection of smoke arising from the head and being scattered through the aether, or from the light of the head'.

Finally, in prop. 41, Bk 3, he explains the fact that the tails of comets ascend from the heads and extend away from the sun by an analogy to the ascent of smoke in a chimney 'by the impulse of the air in which it floats', the air being 'rarefied by heat' so as to gain a 'diminished specific gravity'. He assumes that 'the tail of a comet' may 'ascend from the sun in the same way'. The sun's rays, he argues, will 'warm the reflecting particles' of the medium through which they pass, and these 'reflecting particles, warmed by this action [of the sun's rays], will warm the aethereal aura in which they are entangled'. It is not immediately clear as to whether these reflecting particles compose the comet's tail or are a part of the celestial matter (see D. Gregory's comments in Newton's *Correspondence*, vol. 3, pp. 311, 316, no. 9). The aethereal aura will accordingly become rarefied and have a decrease in the 'specific gravity with which it was formerly tending toward the sun', and so 'that aura will ascend and will carry with it the reflecting particles of which the comet's tail is composed'. Motte (1729) translated Newton's "aethereal aura" by "aethereal air". But Newton had just been using "aer" for "air" in this paragraph, as was his wont elsewhere in the *Principia*. The word "aura" at that time rather meant a subtle or tenuous exhalation, and had been used by Kepler (whose name is mentioned by Newton at the head of this paragraph) as a subtle medium or very fine matter in space. This is, with regard to the sun, very much like the 'aethereal medium' introduced in Newton's hypothesis for 'explaining the Properties of Light' in 1675: 'an aethereal medium much of the same constitution with air, but far rarer, subtler, and more strongly elastic'.

11 Newton to Leibniz, 16 Oct. 1693, *Correspondence*, vol. 3, pp. 286, 287.

12 And, as Newton knew by the time of *De motu* in late 1684, because of the lack of sensible resistance to the motion of comets and planets, which are not slowed down appreciably as they move in their orbits.

13 Yet another proof that there cannot be a "dense" aether is given at the end of Book 2 of the *Principia*, in which Newton shows that Cartesian vortices require conditions which contradict observations concerning the relative orbital speeds of planets in different parts of their orbits.

14 See note 11 *supra*. Newton's statement about a 'subtle matter' is all the more remarkable in that he had just asserted that the heavens have to be 'stripped of all matter', i.e., non-subtle or ordinary matter.

15 See Bopp (1929), Fatio (1949), and Newton's *Correspondence*, vol. 3, p. 69. Fatio's hypothesis was presented to the Royal Society on 27 June 1688. Newton is reported to have said 'that there is but one possible Mechanical cause of Gravity', that which Fatio 'had found out'.

For the text that Newton at one time planned to add to a new edition of the *Principia*, see Hall & Hall (1962), pp. 313, 315. See also Cohen (1971), pp. 184–185.

16 See note 7 *supra*.

17 Whiteside (1976), p. 318.

18 *Correspondence*, vol. 2, p. 360.

19 *Ibid.*, p. 351.

20 See §5 *infra*.

21 At least, he would no longer have firmly believed in the possibility that gravitation might be caused by such an aether (acting in vortices, in a simple shower of aethereal particles, or in relation to a density gradient), although he tried later on to find explanations of universal gravity in a different kind of aether. Such an aether could not have produced mutual gravitation. Further evidence as to the date of the pendulum experiment is given in note 23 *infra*. On the "alchemical" aether of the 1670's, see n.30, §8.

22 See Hall & Hall (1962), pp. 285–286. This scholium, mentioning interplanetary perturbations, was composed in the first months of 1685 (or possibly late 1684).

[23] The statement that 'Aetheris enim puri resistentia quantum sentio vel nulla est vel perquam exiqua' ('For the resistance of pure aether, in my judgment, is either nil or extremely small') actually occurs in the scholium added to the revised *De motu*, most likely written early in 1685, and certainly not before December 1684. This text is printed in Hall & Hall (1962), pp. 261, 286, and in Herivel (1965), pp. 297, 301, C. Since Newton does not here refer to the pendulum experiment in which he compared the oscillatory motion when the pendulum bob was empty (save for air) and when it was filled with metal, I believe we may safely conclude that this experiment (designed to test whether the aether has a measurable resistance) had to have been performed later than December 1684 and probably later than the first months of 1685.

The verb "sentio, sentire", used by Newton, literally means "to perceive by the senses", which might seem to imply that Newton had already performed an experiment, such as the pendulum experiment. But this verb has also the general sense of think, deem, judge, propose. Hence, by the phrase "quantum sentio", Newton would have meant no more than "in my judgment" or "in my opinion" or even "I think". I believe that if Newton had already made an experiment of the sort he described in the *Principia*, he would not have used a phrase such as "quantum sentio" but rather something like "as experiment shows" or "as I have found by experiment". This interpretation of "sentio" would place the date of the experiment later in 1685, during the actual writing of the *Principia*. On the other hand, should the phrase "quantum sentio" have been intended to refer to an experiment, with the meaning of "so far as I am aware by sense perception", then the experiment would presumably have been made in December 1684 or early in 1685, that is, after the first version of *De motu* (November 1684) and before the revision. It should be noted, however, that had Newton intended to indicate "so far as the senses can perceive", he would probably have used the adverb "sensibiliter", as he did in prop. 48, bk 2, of the *Principia*. Hence considering the advanced level of scientific reasoning that lay behind the interpretation of the pendulum experiment, I cannot see any possibility of a date prior to late in 1685 for the performance of this experiment.

There is a similarity between the sailor's not feeling an aether-draft and the pendulum experiment. The sailors do not feel the motion of the aether when they are protected from the wind, for example, when in an airtight cabin. In the pendulum experiment, the metal inside the hollow pendulum bob – "protected" from the air – encounters no sensible resistance from the aether flowing through the box comprising the pendulum bob.

[24] In the concluding scholium to sect. 11, bk 1, Newton lists the possible ways in which there may be 'attraction' (or 'any endeavor of bodies to approach one another'): these are 'the action of bodies either mutually seeking one another or agitating one another by means of spirits emitted' or by 'the action of aether or of air or of any medium whatsoever – whether corporeal or incorporeal – in any way impelling toward one another the bodies floating therein'.

§ 8: CONCLUSION

[1] In relation to particulate forces, Newton actually wrote as follows. In the draft 'Conclusio' for the first edition of the *Principia*, he used such phrases as 'I suspect . . .', 'I do not clearly see why . . . not . . .', 'It is probable that . . .', '. . . but is, I suspect . . .', '. . . if

it shall be settled . . .', 'For I think . . .', 'Whence I suspect . . .', '. . . if only it is possible to prove'; Hall & Hall (1962), pp. 333, 336, 341, 342, 343, 345. In the draft preface, he wrote, similarly, 'For I suspect . . .', 'I therefore propose . . .'; *ibid.*, pp. 307, 308. Cf. 'Hypoth. 2', quoted in Cohen (1966), p. 180, and in Westfall (1971), pp. 379–380, 'The truth of this Hypothesis I assert, not because I cannot prove it, but it very probably . . .'. See, further, Cohen (1980), pp. 255–256, and n. 5 *infra*.

[2] That is, there is no development of the subject in the "Newtonian style"; and no general exposition or development of mathematical laws or principles, proved by the use of mathematical techniques, as is done in the *Principia*.

[3] Westfall (1975), p. 224, and at length in Westfall (1971).

[4] This topic has been explored by Westfall in Westfall (1971) and also in various articles by Westfall, Guerlac, and others. We shall know more concerning the development of the Queries with the publication of Henry Guerlac's edition of the *Opticks*.

[5] This set of five hypotheses is printed *in extenso* in Cohen (1966) from U.L.C. MS Add. 3973, fols. 335–338. The first discusses 'certain spheres of activity' of the 'particles of bodies' within which 'they attract or shun one another'. Here Newton asserts that 'the attractive vertue of the whole magnet is composed of the attractive vertues of all its particles & the like is to be understood of the attractive vertues of electrical & gravitating bodies'. 'Hypoth. 3' states that bodies may move through the 'spheres of activities of bodies' without resistance or promotion of motion other than being 'attracted or repelled to or from the centers of those spheres'. 'Hypoth. 4' declares: 'All solid bodies are [composed of?] atoms.' Then 'Hypoth. 5' (which is cancelled) deals with the 'essential properties of bodies' which 'are not yet fully known to us'.

[6] Newton has crossed out an earlier version, reading: 'Philosophiae naturalis Principia mathematica by such a mathematical way of arguing as has [given satisfaction *crossed out*] procured the assent of all the ablest Mathematicians'.

[7] Newton has crossed out 'but now these things being received by Mathematicians'.

[8] Scholium following prop. 78, Bk 1. Here Newton is not so much concerned with physical particulate units (such as atoms) of which bodies are composed, but rather the 'particles' into which a body may be conceived to have been mathematically divided, as in problems of integration.

Sections 12 & 13 of Book 1 of the *Principia* deal respectively with the attractions of spherical and of non-spherical bodies, that is, with the attractions of gross bodies rather than of mass points. Newton's mathematical mode of analysis is to consider a body as a collection of tiny particles, each of which has the power of attracting according to some law; then to sum up all of these forces (tantamount to performing a mathematical integration), so as to get the total force of a gross body from the sum of the forces of the particles into which Newton conceives the body to be mathematically divided.

[9] See note 23 to §6 *supra*.

[10] Def. 7: this 'measure' is 'proportional to the velocity' which a centripetal force 'generates in a given time'.

[11] As such it is distinct from the 'motive quantity' or 'measure' of a centripetal force: proportional to the momentum ('quantity of motion' as defined in Def. 2) generated 'in a given time'.

[12] Prop. 8, Bk 3.

[13] See Cohen (1980), p. 14.

[14] Stewart (1877), vol. 2, pp. 272–273.

15 For the moon there is also an argument (prop. 3, bk 3) from the 'very slow motion of the moon's apogee'. Of course, the moon – as Newton shows in Book Three – has a complex motion because of the perturbations caused by the sun's gravitational force.

16 This is the famous moon test (see § 3 *supra*).

17 Schol. foll. prop. 5, bk 3.

18 Corol. 1 to prop. 6, bk 3. Corol. 3 states that all planets 'gravitate' toward one another.

19 Westfall (1971), pp. 377–378.

20 *Ibid.*

21 *Ibid.*, p. 265.

22 Westfall (1975). Westfall's extreme point of view of 1975 had been presaged in Westfall (1972).

23 Westfall (1975), pp. 225, 265.

24 First published in Hall & Hall (1962), where attention is called to Newton's belief in the forces acting between particles, without an aether.

25 Westfall (1975), p. 224. In the editor's introduction to the volume containing this essay, William R. Shea states: "Now that we know how deeply fascinated Newton was by alchemical treatises and what store he put in them, we shall have to reconsider, as Westfall suggests, the route that led to the *Principia* themselves.' See Righini Bonelli & Shea (1975), p. 7.

26 Professor Dobbs (1975) has produced a major pioneering study of Newton's early alchemy, based on a careful collation of Newton's alchemical manuscripts and his records of "chymical" (or alchemical) experimentation. This work is the first published full-length investigation of this topic. The author's remarkable command of the literature and techniques of alchemy have won universal respect. But however much we may agree with her authoritative conclusions concerning Newton and alchemy, we do not necessarily have to endorse all of her general conclusions concerning alchemy in relation to the rest of Newton's scientific thought, especially his work in dynamics and celestial mechanics. For example, can one truly describe Newton's research in mathematics and the production of the ideas and techniques of the *Principia*, and much of the work in optics, by saying (p. 230) that 'the whole of his career after 1675 may be seen as one long attempt to integrate alchemy and the mechanical philosophy'? Her book stops at about 1675, and all scholars eagerly look forward to the sequel, now in progress. See the reviews of Dr. Dobbs's book by D. T. Whiteside (1977) in *Isis*, vol. 68, pp. 116–121, and by Karin Figala in 1977 in *History of Science*, vol. 15, pp. 102–137. Dr. Figala has completed a doctoral dissertation (in German) on Newton's alchemy which I hope may soon be published. It is greatly to be wished that she will prepare a full summary of her findings to be published in English.

27 Dobbs (1975), pp. 220, 221; on p. 212 she expresses her belief that 'the concept of alchemical attraction had allowed him to accept the ontological reality of celestial forces through analogy with the alchemical "magnets" he had studied'.

28 In Westfall's *Force in Newton's Physics* (1971), it is suggested that the new concept of force (said to have been developed in 1679/80) was related to the failure of the pendulum experiment (see § 7 *supra*) to reveal the existence of an aether dense enough to be the source of various physical phenomena, including gravity. In Dobbs (1975), p. 230, it is said: 'The original concept of the aether failing to stand further experimental testing, Newton abandoned it for a new concept of force which in fact [!] incorporated

the alchemical active principles into mechanism in a radically new way'. Westfall (1971), pp. 375–376, says of this experiment that Newton 'did not indicate the exact date'. Nevertheless, he concludes, 'It is not unreasonable to speculate that the experiment was connected with the reversal of opinion that I have tentatively placed about 1679'. But on p. 378, Westfall more cautiously argues that 'whether or not' Newton's 'change in outlook' occurred 'in 1679, it had occurred by 1686–7 when, as he completed the *Principia*, Newton composed his next extensive speculations on the nature of physical reality'.

Westfall (1975), pp. 223–224, declares that 'Some years before the *Principia*, according to his [Newton's] own account, he had convinced himself by an experiment with a pendulum that the aether does not exist'. On the problem of dating of this experiment, see §7 *infra*. Prof. Dobbs would see her own study of Newton's alchemy pointing 'rather plainly to a specific episode which might well have predisposed Newton to think in terms of attractive forces: the alchemical success he reported in the "Clavis" of Keynes MS 18. When his aether failed him, he perhaps remembered that very special mercury which the attractive powers of the "magnet" had given him'. See Dobbs (1975), p. 212.

29 This point was the topic of a very active scholarly discussion at the Cambridge meetings (Churchill College, Aug. 1977), in relation to the possibility of a Hermetic or alchemical source of the Newtonian concept of universal gravity.

30 Westfall himself (see n. 28 *supra*) agrees that Newton's change of mind with respect to particulate forces may have occurred only by 1686–7, too late – in that case – to have influenced the formulation of the mutually acting force of universal gravity.

31 Westfall (1971), p. 377.

32 See §6 *supra*.

33 See §6 *supra*.

34 Newton had first discussed electrical 'effluvia' in his hypothesis on the aether in 1675; see Cohen & Schofield (1956), p. 180, where he linked electrical attraction and gravity.

35 Scholium at the end of sect. 11, Bk 1, *Principia*.

36 In the preface to the second edition of the *Principia*, Cotes argued that gravitational force should be given the status of an independent primary concept: 'Among the primary qualities of all bodies universally, either gravity will have a place, or extension, mobility, and impenetrability will not.' Newton, however, never agreed with this position. In his correspondence with Bentley in 1691, he specifically said that he did not consider gravity to be 'essential and inherent to matter'.

37 See Westfall (1971), pp. 391–395.

38 Newton to Bentley, 25 Feb. 1692/3; *Correspondence*, vol. 3, pp. 253–54. As Westfall points out (1971), p. 396, Newton had 'not left it to the consideration of his readers', since Book Two of the *Principia* contained many arguments to show that there can be no material aether. What, then, is an "immaterial" aether? Westfall suggests that for Newton this was 'the infinite omnipotent God, who by His infinity constitutes absolute space and by His omnipotence is actively present throughout it'.

39 Newton's *Correspondence*, vol. 3, p. 70.

40 See also More (1941), Ihde (1964), Boas (1958), Maddison (1969). For the older literature, see the extensive bibliography in Dobbs (1975).

41 There are also valuable, and shorter, earlier studies, as those by F. Sherwood Taylor,

D. Geoghegan, M.S. Churchill; and also studies of Newton's chemistry and theory of matter, as by A. Rupert Hall & Marie Boas Hall.

42 See, e.g., Hill (1965). Important, too, in this regard are the studies of P. M. Rattansi (1972) and of J. E. McGuire (1967, 1968); and especially the work of McGuire and Rattansi (1966) on the Priscan tradition.

43 See §4 *supra.*

44 This has emerged particularly from the investigations reported in Rattansi (1972).

45 Newton's hypothesis on the aether was printed in Birch (1957), vol. 3, pp. 248–260; reproduced in Cohen & Schofield (1958), pp. 178–190, 250–254. For an analysis of the alchemical aspects of this aether, see Dobbs (1975).

46 In particular, the researches of B. J. T. Dobbs and of Karin Figala have shown how Newton's ideas about the structure of matter were closely related to his alchemical pursuits.

One of the most significant aspects of the research by Prof. Dobbs (1975) is that there is a correlation between actual experiment (whether chemical or alchemical) and Newton's reading or studying of texts; see, also, Westfall (1975). That is, Newton was not a pure armchair Hermetic philosopher but a chemical alchemist.

47 As has been done by Prof. Dobbs in her contribution to the Newtonian symposium in Cambridge in 1977.

48 See Aiton (1969); Newton made this claim in a letter to Halley, 20 June 1686, *Correspondence*, vol. 2, p. 436. Of course, any power that spreads out uniformly in all directions from a body will diminish as the square of the distance.

49 The usual claims for the influence of alchemy or the Hermetic philosophy do not include the quantitative formulation – neither the direct proportionality of the gravitational force to the product of the masses, nor its inverse proportionality to the square of the distance.

50 This mathematical treatment is summarized in words in the *Opticks*, prop. 10, Bk 2, pt. 3.

51 See his forthcoming edition of the *Lectiones opticae.*

52 Chiefly Lohne (1961), esp. p. 397.

53 See especially Bechler (1973).

54 That is, no one has as yet systematically distinguished between genuine measurements or conclusions based upon careful genuine experiments and the alleged experimental results that may in fact have been conclusions from mathematical or other theories or models. Such a distinction will be easier to make once we have available the edition of the *Lectiones opticae* and of Newton's optical MSS currently being undertaken by Alan Shapiro and the edition of the *Opticks* being prepared by Henry Guerlac. It was obvious that Newton had used mathematical techniques in obtaining some of his results. Hence, in the posthumous 'fourth edition, corrected' (London 1730), the editor added reference to the more mathematical posthumously published *Lectiones opticae*. For instance, in computing the results of an experiment in prop. 7, bk 1, pt 1 (p. 95 of Newton (1952), *Opticks*), he gave a result in which a certain diameter had the value $(R^2/I^2) \times (S^3/D^2)$ 'very nearly, as I gather by computing the Errors of the Rays by the Method of infinite Series, and rejecting the Terms, whose Quantities are inconsiderable'. The editor adds: '*How to do this, is shewn in our* Author's Lect. Optic. *Part* I. *Sect*. IV. *Prop*. 31.'

55 In the *Opticks*, Newton does not make very much use of the preliminary axioms, since the propositions are not developed by logical or mathematical analysis or techniques

but are proved, as Newton says, 'by Experiment'. The *Opticks* does not make use of mathematics in the sense of algebra, fluxions, or infinite series, but does use simple or elementary geometry (but not the theory of limits applied geometrically, as in the *Principia*), and it is "mathematical" in the sense of being quantitative (i.e., using numbers actually and allegedly deriving from direct measurement).

It has recently been argued, in the *Dictionary of the History of Ideas*, that the *Opticks* 'can hardly be called nonmathematical, although little more than some simple geometry and arithmetic is needed to understand it. In spirit it is as good an example of Newton's "mathematical way" as the *Principia*: light is treated as a mathematical entity, as *rays* that can be represented by lines; the axioms with which he begins are the accepted laws of optics; and numbers – the different refrangibilities – serve as precise tags to distinguish the rays of different colors and to compare their behavior in reflection, refraction, and diffraction. Wherever appropriate, and this is most of the time, his language of experimental description is the language of number and measure. It is this which gives Newton's experiments their particular cogency.' In this article, the characterization of mathematics goes contrary to the strict sense of mathematical techniques in what I have called the "Newtonian style". The phrase used in the above quotation, 'Newton's "mathematical way" ', of which the *Opticks* is said to be a 'good . . . example', comes from the title of a paper by Strong (1951); it is also the title of the section of the article from which the above quotation is taken: 'Newton's mathematical way'. This phrase ('mathematical way') comes from the English translation (1728) of the first version of Newton's *System of the world*, a phrase which was gratuitously inserted by the translator and for which there is no corresponding phrase in any of the copies of Newton's original Latin text; see Dundon (1969), Cohen (1969).

56 For the simple reason that all such attempted exercises by Newton had been failures. As a result, he wrote up some of his optical researches in a more popular form and published them in 1704 as the *Opticks*.

57 According to Lohne (1961), p. 399, 'If his dreams had come true, the mathematical science of colors would have eclipsed even his *Principia*.'

58 That is, the notion that, whatever their cause, there might be solar-planetary forces could have been reinforced or made plausible by a cognizance of alchemical forces along with those of magnetism, electricity, and terrestrial weight.

59 Possibly, a case can be made for the original Hypoth. 4 at the beginning of Bk 3 in the first edition (1687). McGuire (1967) alleges that this was a 'transmutation' hypothesis, and so had alchemical implications. In any event, I have shown in Cohen (1964) that both Newton and the Newtonians considered this to have been the 'hypothesis' of Descartes and the Aristotelians!

BIBLIOGRAPHY

Aiton, E. J.: 1964, 'The inverse problem of central forces', *Annals of Science* **20**, 81–99.

Aiton, E. J.: 1969, 'Newton's aether-stream hypothesis and the inverse square law of gravitation', *Annals of Science* **25**, 255–260.

Aiton, E. J.: 1969*a*, 'Kepler's second law of planetary motion', *Isis* **60**, 75–90.

Aiton, E. J.: 1972, *The Vortex Theory of Planetary Motions*, London: Macdonald; New York: American Elsevier.

[Babson Collection]: 1950, *A Descriptive Catalogue of the Grace K. Babson Collection of the Works of Sir Isaac Newton, and the Material Relating to Him in the Babson Institute Library, Babson Park, Mass.*, With an introduction by Roger Babson Webber, New York: Herbert Reichner.

[Babson Collection] Supplement: 1955, compiled by Henry P. Macomber, Babson Park (Mass.): The Babson Institute.

Baily, Francis: 1835, *An Account of the Revd. John Flamsteed, the First Astronomer-Royal; Compiled from His Own Manuscripts, and Other Authentic Documents, Never before Published. To Which Is Added, His British Catalogue of Stars, Corrected and Enlarged*, London: printed by the order of the Lords Commissioners of the Admiralty.

Baily, Francis: 1837, *Supplement to the Account of the Revd. John Flamsteed, the First Astronomer-Royal*, London: printed for distribution amongst those persons and institutions only, to whom the original work was presented. Reprint (1966), London: Dawsons of Pall Mall.

Ball, W. W. Rouse: 1893, *An Essay on Newton's "Principia"*, London, New York: Macmillan and Co. Reprint (1972), New York, London: Johnson Reprint Corporation, with an introduction by I. B. Cohen.

Bechler, Zev: 1973, 'Newton's search for a mechanistic model of colour dispersion: a suggested interpretation', *Archive for History of Exact Sciences* **11**, 1–37.

Bechler, Zev: 1974, 'Newton's 1672 optical controversies: a study in the grammar of scientific dissent', pp. 115–142 of Y. Elkana, ed., *The Interaction between Science and Philosophy*, Atlantic Highlands (N.J.): The Humanities Press.

Bechler, Zev: 1974*a*, 'Newton's law of forces which are inversely as the mass: a suggested interpretation of his later efforts to normalise a mechanistic model of optical dispersion', *Centaurus* **18**, 184–222.

Bechler, Zev: 1975, ' "A less agreeable matter": the disagreeable case of Newton and achromatic refraction', *British Journal for the History of Science* **8**, 101–126.

Birch, Thomas: 1756–7, *The History of the Royal Society of London, for the Improving of Natural Knowledge, from Its First Rise. In Which the Most Considerable of Those Papers Communicated to the Society, Which Have Hitherto Not Been Published, Are Inserted in Their Proper Order, as a Supplement to the Philosophical Transactions*, 4 vols., London: printed for A. Millar. Reprint (1968), Brussels: Editions "Culture et Civilisation"; another reprint (1968), New York, London: Johnson Reprint Corporation, with introduction, analyses, and supplementary entries by A. R. Hall.

Boas, Marie: 1958, *Robert Boyle and Seventeenth-Century Chemistry*, Cambridge: at the University Press.

Bopp, K.: 1929, 'Die wiederaufgefundene Abhandlung von Fatio de Duillier: De la cause de la pesanteur', *Schriften der Strassburger Wissenschaftlichen Gesellschaft in Heidelberg*, neue Folge, 10. Heft, 19–66.

Brasch, F. E. (ed.): 1928, *Sir Isaac Newton, 1727–1927, A Bicentenary Evaluation of His Work: A Series of Papers Prepared under the Auspices of the History of Science Society*, Baltimore: The Williams & Wilkins Company.

Buchdahl, Gerd: 1970, 'History of science and criteria of choice', *Minnesota Studies in the Philosophy of Science*, vol. 5 ('Historical and Philosophical Perspectives of Science', ed. by Roger H. Stuewer), pp. 204–245.

Cajori, Florian: 1928, 'Newton's twenty years' delay in announcing the law of gravitation', pp. 127–188 of Brasch (ed.) (1928).

Chandler, Philip P., II: 1975, *Newton and Clairaut on the Motion of the Lunar Apse.*, San Diego (Cal.): [doctoral dissertation, University of California].

Cohen, I. Bernard: 1946, 'Authenticity of scientific anecdotes', *Nature* **157**, 196–197.

Cohen, I. Bernard: 1953, Review of facsimile edition of Newton's *Principia, Isis* **44**, 287–8.

Cohen, I. Bernard: 1956, *Franklin and Newton: An Inquiry into Speculative Newtonian Experimental Science and Franklin's Work in Electricity as an Example Thereof*, Philadelphia: The American Philosophical Society. Reprint (1966), Cambridge (Mass.): Harvard University Press.

Cohen, I. Bernard: 1960, 'Newton in the light of recent scholarship', *Isis* **51**, 489–514.

Cohen, I. Bernard: 1962, 'The first English version of Newton's "Hypotheses non fingo" ', *Isis* **53**, 379–388.

Cohen, I. Bernard: 1963, 'Pemberton's translation of Newton's *Principia*, with notes on Motte's translation', *Isis* **54**, 319–351.

Cohen, I. Bernard: 1964, ' "Quantum in se est": Newton's concept of inertia in relation to Descartes and Lucretius', *Notes and Records of the Royal Society of London* **19**, 131–155.

Cohen, I. Bernard: 1966, 'Hypotheses in Newton's philosophy', *Physics, Rivista Internazionale di Storia della Scienza* 8, 163–184. Reprinted (1969) in *Boston Studies in the Philosophy of Science*, vol. V, pp. 304–326.

Cohen, I. Bernard: 1967, 'Newton's attribution of the first two laws of motion to Galileo', *Atti del Symposium Internazionale di Storia, Metodologia, Logica e Filosofia della Scienza 'Galileo nella Storia e nella Filosofia della Scienza'*, pp. xxv–xliv. Collection des Travaux de l'Académie Internationale d'Histoire des Sciences. N. 16. Vinvi (Firenze): Gruppo Italiano di Storia della Scienza.

Cohen, I. Bernard: 1967*a*, 'Newton's second law and the concept of force in the *Principia*', pp. 143–185 of Palter (ed.) (1970).

Cohen, I. Bernard: 1969, 'Newton's *System of the World*: some textual and bibliographical notes', *Physis* **11**, 152–166.

Cohen, I. Bernard: 1969*a*, Introduction, pp. vii–xxii, to *A Treatise of the System of the World* by Sir Isaac Newton [a facsimile reprint of the second edition, 1731], London: Dawsons of Pall Mall.

Cohen, I. Bernard: 1971, *Introduction to Newton's Principia*, Cambridge (Mass.): Harvard University Press; Cambridge (England): at the University Press. Revised reprint (1978), in paperback.

Cohen, I. Bernard: 1974, 'Newton's theory vs. Kepler's theory and Galileo's theory', pp. 299–338 of Y. Elkana (ed.): *The Interaction between Science and Philosophy*, Atlantic Highlands (N.J.): Humanities Press.

Cohen, I. Bernard: 1974*a*, 'Newton, Isaac [1642–1727]', *Dictionary of Scientific Biography*, edited by Charles C. Gillispie, vol. 10, pp. 42–103, New York: Charles Scribner's Sons.

Cohen, I. Bernard: 1975, *Isaac Newton's Theory of the Moon's Motion (1702)*, Folkestone (Kent): Dawson.

Cohen, I. Bernard: 1976, 'Newtonian astronomy: the steps toward universal gravitation', *Vistas in Astronomy* **20**, 85–98.

Cohen, I. Bernard: 1976*a*, 'The eighteenth-century origins of the concept of scientific revolution', *Journal of the History of Ideas* **37**, 257–288.

Cohen, I. Bernard: 1977, 'The many faces of the history of science', pp. 65–110 of Charles F. Delzell (ed.): *The Future of History*, Nashville (Tenn.): Vanderbilt University Press.

Cohen, I. Bernard: 1980, *The Newtonian Revolution, with Illustrations of the Transformation of Scientific Ideas*, Cambridge (England): Cambridge University Press.

Cohen, I. Bernard: 1981, 'Newton's discovery of gravity', *Scientific American* **244**, no. 3 (March), 166–179.

Cohen, I. Bernard: 1982, *Revolution in Science: History, Analysis, and Significance of a Name and a Concept*, Cambridge (Mass.): Harvard University Press [in press].

Cohen, I. Bernard and Robert E. Schofield (eds.): 1958, *Isaac Newton's Papers and Letters on Natural Philosophy and Related Documents*, Cambridge (Mass.): Harvard University Press; Cambridge (England): at the University Press. Revised and expanded second edition (1978), Cambridge (Mass.), London (England): Harvard University Press.

Collins, John, et al.: 1856, *Commercium epistolicum J. Collins et aliorum de analysi promota, etc., ou, Correspondance de J. Collins et d'autres savants célèbres du XVIIe siècle, relative à l'analyse supérieure, réimprimée sur l'édition originale de 1712 avec l'indication des variantes de l'édition de 1722, complétée par une collection de pièces justificatives et de documents, et publiée par J.-B. Biot et F. Lefort*, Paris: Mallet-Bacheller.

De Beer, G. R. *See* McKie and De Beer (1952).

Debus, Allen G. (ed.): 1972, *Science, Medicine, and Society in the Renaissance: Essays to Honor Walter Pagel*, 2 vols., New York: Science History Publications (a division of Neale Watson Academic Publications).

Debus, Allen G.: 1978, 'The pseudo sciences and the history of science', *The University of Chicago Library Society Bulletin* **3**, 3–20.

Dobbs, Betty Jo Teeter: 1975, *The Foundations of Newton's Alchemy, or "The Hunting of the Greene Lion"*, Cambridge (England), London, New York, Melbourne: Cambridge University Press.

Dugas, René: 1950, *Histoire de la mécanique*, Neuchâtel: Éditions du Griffon. Translated by J. R. Maddox as: *A History of Mechanics*, Neuchâtel: Éditions du Griffon; New York: Central Book Company, 1955.

Dugas, René: 1954, *La mécanique au XVIIe siècle (des antécédents scolastiques à la pensée classique)*, Neuchâtel: Éditions du Griffon. Translated by Freda Jacquot as: *Mechanics in the Seventeenth Century (from the Scholastic Antecedents to Classical Thought)*, Neuchâtel: Éditions du Griffon; New York: Central Book Company, 1958.

Duhem, Pierre: 1954, *The Aim and Structure of Physical Theory*, trans. from the French by Philip P. Wiener, Princeton: Princeton University Press.

Dundon, Stanislaus: 1969, 'Newton's "mathematical way" in the *De mundi systemate*', *Physis* **11**, 195–204.

Edelston, J.: 1850, *Correspondence of Sir Isaac Newton and Professor Cotes, Including Letters of Other Eminent Men, Now First Published from the Originals in the Library of Trinity College, Cambridge; together with an Appendix, Containing Other Unpublished Letters and Papers by Newton*, London: John W. Parker; Cambridge (England): John Deighton.

Fatio De Duillier, Nicolas: 1949, 'De la cause de la pesanteur. Mémoire présenté à la Royal Society le 26 février 1690. [Reconstitué et publié avec une introduction par Bernard Gagnebin]', *Notes and Records of the Royal Society of London* **6**, 105–160.

Figala, Karin: 1977, *Die "Kompositionshierarchie" der Materie – Newtons quantitative Theorie und Interpretation der qualitativen Alchemie*, Munich: [unpublished *Habilitationsschrift* in the Technische Universität, München].

Gagnebin, Bernard. *See* Fatio De Duillier (1949).

Gray, George J.: 1907, *A Bibliography of the Works of Sir Isaac Newton, together with a List of Books Illustrating His Works*, Second edition, revised and enlarged, Cambridge (England): Bowes and Bowes.

Greenstreet, W. J. (ed.): 1927, *Isaac Newton, 1642–1727. A Memorial Volume Edited for the Mathematical Association*, London: G. Bell and Sons.

Grigorian, Achote T.: 1964, 'Les études newtoniennes de A. N. Krylov', pp. 198–207 of I. B. Cohen and R. Taton (eds.), *Mélanges Alexandre Koyré: vol. 1, L'Aventure de la science*, Paris: Hermann (Histoire de la Pensée, 12).

Guerlac, Henry: 1963, *Newton et Epicure*, Paris: Palais de la Découverte [Histoire des Sciences: D-91].

Guerlac, Henry: 1963*a*, 'Francis Hauksbee: expérimentateur au profit de Newton', *Archives Internationales d'Histoire des Sciences* **16**, 113–128.

Guerlac, Henry: 1964, 'Sir Isaac and the ingenious Mr. Hauksbee', pp. 228–253 of I. B. Cohen and R. Taton (eds.), *Mélanges Alexandre Koyré*: vol. 1, *L'Aventure de la science*, Paris: Hermann (Histoire de la Pensée, 12).

Guerlac, Henry: 1967, 'Newton's optical aether. His draft of a proposed addition to his *Opticks*', *Notes and Records of the Royal Society of London* **22**, 45–57.

Guerlac, Henry: 1972, 'Hauksbee, Francis [1666–1713]', *Dictionary of Scientific Biography*, edited by Charles C. Gillispie, vol. 6, pp. 169–176, New York: Charles Scribner's Sons.

Guerlac, Henry: 1973, 'Newton and the method of analysis', *Dictionary of the History of Ideas*, edited by Philip P. Wiener, vol. 3, pp. 378–391, New York: Charles Scribner's Sons.

Guerlac, Henry: 1977, *Essays and Papers in the History of Modern Science*, Baltimore: The Johns Hopkins Press.

Hall, A. R.: 1957, 'Newton on the calculation of central forces', *Annals of Science* **13**, 62–71.

Hall, A. R.: 1963, 'The date of "On motion in ellipses" ', *Archives Internationales d'Histoire des Sciences* **16**, 23–28.

Hall, A. R. and Marie Boas Hall: 1959, 'Newton's electric spirit: four oddities', *Isis* **50**, 473–476.

Hall, A. R. and Marie Boas Hall: 1959*a*, 'Newton's "mechanical principles" ', *Journal of the History of Ideas* **20**, 167–178.

Hall, A. R. and Marie Boas Hall: 1960, 'Newton's theory of matter', *Isis* **51**, 131–144.

Hall, A. R. and Marie Boas Hall (eds.): 1962, *Unpublished Scientific Papers of Isaac Newton. A Selection from the Portsmouth Collection in the University Library, Cambridge*, Cambridge (England): at the University Press.

Herivel, J. W.: 1960, 'Newton's discovery of the law of centrifugal force', *Isis* **51**, 546–553.

Herivel, J. W.: 1960*a*, 'On the date of composition of the first version of Newton's tract *De motu*', *Archives Internationales d'Histoire des Sciences* 13, 67–70.

Herivel, J. W.: 1961, 'Interpretation of an early Newton manuscript', *Isis* 52, 410–416.

Herivel, J. W.: 1965, *The Background to Newton's Principia. A Study of Newton's Dynamical Researches in the Years 1664–84*, Oxford: at the Clarendon Press.

Herivel, J. W.: 1965*a*, 'Newton's First Solution to the Problem of Kepler Motion', *British Journal for the History of Science* 2, 350–354.

Hill, Christopher: 1965, *Intellectual Origins of the English Revolution*, London: Oxford University Press. Reprint (1972), London: Panther Books.

Husserl, Edmund: 1970, *The Crisis of European Sciences and Transcendental Phenomenology: An Introduction to Phenomenological Philosophy*, translated, with an introduction by David Carr, Evanston (Illinois): Northwestern University Press. The German original (1954) was published under the title: *Die Krisis der europäischen Wissenschaften und die transcendentale Phänomenologie: eine Einleitung in die phänomenologische Philosophie*, The Hague: Martinus Nijhoff.

Huygens, Christiaan: 1888–1950, *Oeuvres complètes de Christiaan Huygens*, publiées par la Société Hollandaise des Sciences, 22 volumes, La Haye: Martinus Nijhoff.

Ihde, Aaron J.: 1964, 'Alchemy in reverse: Robert Boyle on the degradation of gold', *Chymia* 9, 47–57.

Jouguet, E.: 1924, *Lectures de mécanique: la mécanique enseignée par les auteurs originaux*, nouveau tirage avec notes et additions, 2 vols., Paris: Gauthier-Villars et Cie.

King, Lord: 1829, *The Life of John Locke, with Extracts from his Correspondence, Journals, and Common-place Books*, London: Henry Colburn. Augmented edition (1830), 2 vols, London: Henry Colburn and Richard Bentley. Reprint (1972), New York: Burt Franklin.

Koyré, Alexandre: 1952, 'An unpublished letter of Robert Hooke to Isaac Newton', *Isis* 43, 312–337.

Koyré, Alexandre: 1955, 'Pour une édition critique des oeuvres de Newton', *Revue d'Histoire des Sciences* 8, 19–37.

Koyré, Alexandre: 1956, 'L'Hypothèse et l'expérience chez Newton', *Bulletin de la Société française de Philosophie* 50, 59–79.

Koyré, Alexandre: 1960, 'Les Regulae philosophandi', *Archives Internationales d'Histoire des Sciences* 13, 3–14.

Koyré, Alexandre: 1965, *Newtonian Studies*, Cambridge (Mass.): Harvard University Press; London: Chapman & Hall. More than half the volume consists of a previously unpublished study on 'Newton and Descartes', pp. 53–200; there are also reprints of Koyré (1952), Koyré (1956) in English trans., Koyré (1960) in English trans., and other studies.

Koyré, Alexandre and I. Bernard Cohen: 1960, 'Newton's "electric & elastic spirit" ', *Isis* 51, 337.

Kriloff, A. N.: 1924, 'On a theorem of Sir Isaac Newton', *Monthly Notices of the Royal Astronomical Society* 84, 392–395.

Lohne, Johannes A.: 1960, 'Hooke versus Newton. An analysis of the documents in the case of free fall and planetary motion', *Centaurus* 7, 6–52.

Lohne, Johannes A.: 1961, 'Newton's "proof" of the sine law', *Archive for History of Exact Sciences* 1, 389–405.

Lohne, Johannes A.: 1965, 'Isaac Newton: the rise of a scientist 1661–1671', *Notes and Records of the Royal Society of London* **20**, 125–139.

Lohne, Johannes A.: 1967, 'The increasing corruption of Newton's diagrams', *History of Science* **6**, 69–89.

Lohne, Johannes A.: 1968, 'Experimentum crucis', *Notes and Records of the Royal Society of London* **23**, 169–199.

Lohne, Johannes A. and Bernhard Sticker: 1969, *Newtons Theorie der Prismenfarben mit Übersetzung und Erläuterung der Abhandlung von 1672*, Munich: Werner Fritsch.

Mach, Ernst: 1960, *The Science of Mechanics, a Critical and Historical Account of Its Development*, translated by Thomas J. McCormack, new introduction by Karl Menger. Sixth edition, with revisions through the ninth German edition, La Salle (Illinois): The Open Court Publishing Company.

Maddison, R. E. W.: 1969, *The Life of the Honourable Robert Boyle, F.R.S.*, London: Taylor & Francis; New York: Barnes and Noble.

Maeyama, Yasukatsu: 1971, *Hypothesen zur Planeten-theorie des 17. Jahrhunderts*, Frankfurt am Main: Institut für Geschichte der Naturwissenschaften, Johann Wolfgang Goethe-Universität.

Manuel, Frank E.: 1968, *A Portrait of Isaac Newton*, Cambridge (Mass.): The Belknap Press of Harvard University Press.

McGuire, J. E.: 1966, 'Body and void and Newton's *De mundi systemate*: some new sources', *Archive for History of Exact Sciences* **3**, 206–248.

McGuire, J. E.: 1967, 'Transmutation and immutability: Newton's doctrine of physical qualities', *Ambix* **14**, 69–95.

McGuire, J. E.: 1968, 'The origin of Newton's doctrine of essential qualities', *Centaurus* **12**, 233–260.

McGuire, J. E.: *See also* Westman and McGuire (1977).

McGuire, J. E. and P. M. Rattansi: 1966, 'Newton and the "pipes of Pan" ', *Notes and Records of the Royal Society of London* **21**, 108–143.

McKie, Douglas and G. R. De Beer: 1952, 'Newton's apple', *Notes and Records of the Royal Society of London* **9**, 46–54, 333–335.

Mercator, Nicolaus: 1696, *Institutionum astronomicarum libri duo* . . . , London: typis Gulielmi Godbid, sumptibus Samuelis Simpson, bibliopolae Cantabrigiensis.

More, Louis Trenchard: 1941, 'Boyle as alchemist', *Journal of the History of Ideas* **2**, 61–76.

Newton, Isaac: 1715, '[Recensio libri=] An account of the book entitled *Commercium epistolicum Collinii & aliorum, De analysi promota*, published by order of the Royal Society, in relation to the dispute between Mr Leibnits and Dr Keill, about the right of invention of the new geometry of fluxions, otherwise call'd the differential method', *Philosophical Transactions* **29**, 173–224.

Newton, Isaac: 1728, *A Treatise of the System of the World*, translated into English, London: printed for F. Fayram.

Newton, Isaac: 1728*a*, *De mundi systemate liber*, London: impensis J. Tonson, J. Osborn & T. Longman.

Newton, Isaac: 1728*b*, *Optical Lectures Read in the Publick Schools of the University of Cambridge, Anno Domini, 1669*. By the late Sir Isaac Newton, then Lucasian Professor of the Mathematicks. Never before printed. Translated into English out of the original Latin. London: printed for Francis Fayram.

Newton, Isaac: 1729, *The Mathematical Principles of Natural Philosophy*, translated into English by Andrew Motte, *To Which Are Added, the Laws of the Moon's Motion, According to Gravity*, by John Machin, in two volumes, London: printed for Benjamin Motte. Reprint (1968), with an introduction by I. Bernard Cohen. London: Dawsons of Pall Mall.

Newton, Isaac: 1934, *Sir Isaac Newton's Mathematical Principles of Natural Philosophy and His System of the World*, translated into English by Andrew Motte in 1729, the translations revised, and supplied with an historical and explanatory appendix, by Florian Cajori, Berkeley: University of California Press.

Newton, Isaac: 1952, *Opticks or a Treatise of the Reflections, Refractions, Inflections and Colours of Light*, based on the fourth edition: London, 1730, with a Foreword by Albert Einstein; an Introduction by Sir Edmund Whittaker; a Preface by I. Bernard Cohen; and an analytical table of contents prepared by Duane H. D. Roller, New York: Dover Publications.

Newton, Isaac: 1959–1977, *The Correspondence of Isaac Newton*, Vol. 1, 1661–1675, Vol. 2, 1676–1687, Vol. 3, 1688–1694, edited by H. W. Turnbull; Vol. 4, 1694–1709, edited by J. F. Scott; Vol. 5, 1709–1713, Vol. 6, 1713–1718, Vol. 7, 1718–1727, edited by A. Rupert Hall and Laura Tilling, Cambridge (England): at the University Press (published for the Royal Society).

Newton, Isaac: 1964, *The Mathematical Works of Isaac Newton*, assembled with an introduction by Derek T. Whiteside, 2 vols., New York, London: Johnson Reprint Corporation (The Sources of Science).

Newton, Isaac: 1967–, *The Mathematical Papers of Isaac Newton*, Vol. 1, 1664–1666, Vol. 2, 1667–1670, Vol. 3, 1670–1673, Vol. 4, 1674–1684, Vol. 5, 1683–1684, Vol. 6, 1684–1691, Vol. 7, 1691–1695, edited by D. T. Whiteside, Cambridge (England): at the University Press, to be completed in 8 volumes.

Newton, Isaac: 1972, *Isaac Newton's Philosophiae naturalis principia mathematica, The Third Edition (1726) with Variant Readings*, edited by Alexandre Koyré, I. Bernard Cohen & Anne Whitman, 2 vols., Cambridge (Mass.): Harvard University Press; Cambridge (England): at the University Press.

Palter, Robert (ed.): 1970, *The Annus Mirabilis of Sir Isaac Newton, 1666–1966*, Cambridge (Mass.), London: M.I.T. Press.

Patterson, Louise Diehl: 1949, 1950, 'Hooke's gravitation theory and its influence on Newton. I: Hooke's gravitation theory. II: The insufficiency of the traditional estimate', *Isis* **40**, 327–341; **41**, 32–45.

Pemberton, Henry: 1728, *A View of Sir Isaac Newton's Philosophy*, London: S. Palmer. Reprint (1972), New York, London: Johnson Reprint Corporation. With an introduction by I. B. Cohen.

Pelseneer, Jean: 1937, 'La pomme de Newton', *Ciel et Terre* **53**, 190–193.

Pighetti, Clelia: 1960, 'Cinquant' anni di studi newtoniani (1908–1959)', *Rivista Critica di Storia della Filosofia*, fascicoli 2–3, pp. 181–203, 295–318.

Popper, Karl R.: 1957, 'The aim of science', *Ratio* **1**, 24–35. Reprinted on pp. 191–205 of Popper (1972).

Popper, Karl R.: 1972, *Objective Knowledge, an Evolutionary Approach*, Oxford: at the Clarendon Press.

[Portsmouth Collection]: 1888, *A Catalogue of the Portsmouth Collection of Books and Papers Written by or Belonging to Sir Isaac Newton, the Scientific Portion of Which*

Has Been Presented by the Earl of Portsmouth to the University of Cambridge, prepared by H. R. Luard, G. G. Stokes, J. C. Adams, and G. D. Liveing, Cambridge: at the University Press.

Rattansi, P. M.: 1972, 'Newton's alchemical studies', pp. 167–182 of Debus (1972), vol. 2.

Rattansi, P. M.: *See also* McGuire and Rattansi (1966).

Rigaud, Stephen Peter: 1838, *Historical Essay on the First Publication of Sir Isaac Newton's Principia*, Oxford: at the University Press. Reprint (1972), New York, London: Johnson Reprint Corporation, with an introduction by I. B. Cohen.

Righini Bonelli, M. L. and William R. Shea: 1975, *Reason, Experiment and Mysticism in the Scientific Revolution*, New York: Science History Publications.

Russell, J. L.: 1964, 'Kepler's laws of planetary motion: 1609–1666', *The British Journal for the History of Science* **2**, 1–24.

Scott, J. F.: *See* Newton (1959–1977).

Stewart, Dugald: 1877, *Elements of the Philosophy of the Human Mind*, 2 vols. Edinburgh: T. and T. Clark [vols. 2 & 3 of *The Collected Works of Dugald Stewart* . . . , edited by Sir William Hamilton].

Streete, Thomas: 1661, *Astronomia Carolina: a New Theorie of the Celestial Motions*, London: for Lodowick Lloyd.

Strong, E. W.: 1951, 'Newton's "mathematical way" ', *Journal of the History of Ideas* **12**, 90–110. Reprint (1957), pp. 412–432 of Philip P. Wiener & Aaron Noland (eds.), *Roots of Scientific Thought*, New York: Basic Books.

Stukeley, William: 1936, *Memoirs of Sir Isaac Newton's Life . . . 1752, Being some Account of His Family and Chiefly of the Junior Part of His Life*, edited by A. Hastings White, London: Taylor & Francis.

Truesdell, C.: 1960, 'A program toward rediscovering the rational mechanics of the Age of Reason', *Archive for History of Exact Sciences* **1**, 3–36.

Truesdell, C.: 1968, *Essays in the History of Mechanics*, New York: Springer-Verlag. Contains Truesdell (1960) and Truesdell (1970).

Truesdell, C.: 1970, 'Reactions of late baroque mechanics to success, conjecture, error, and failure in Newton's *Principia*', pp. 192–232 of *Palter* (1970).

Turnbull, H. W.: 1953, 'Isaac Newton's letters', *Manchester Guardian*, Saturday, 3 Oct. 1953, p. 4. Reprinted in the *Manchester Guardian Weekly*, Thursday, 8 Oct. 1953, p. 11.

Waff, Craig: 1975, *Universal Gravitation and the Motion of the Moon's Apogee: The Establishment and Reception of Newton's Inverse-square Law, 1687–1794*, Baltimore: [unpublished doctoral dissertation, The Johns Hopkins University].

Waff, Craig: 1976, 'Isaac Newton, the motion of the lunar apogee, and the establishment of the inverse-square law', *Vistas in Astronomy* **20**, 99–104.

Wallis, Peter and Ruth Wallis: 1977, *Newton and Newtoniana 1672–1975, a Bibliography*, Folkestone (Kent): Dawson.

Westfall, Richard S.: 1967, 'Hooke and the law of universal gravitation: a reappraisal of a reappraisal', *British Journal for the History of Science* **3**, 245–261.

Westfall, Richard S.: 1969, 'A note on Newton's demonstration of motion in ellipses', *Archives Internationales d'Histoire des Sciences* **22**, 51–60.

Westfall, Richard S.: 1970, 'Uneasily fitful reflections on fits of easy transmission', pp. 88–104 of Palter (1970).

Westfall, Richard S.: 1971, *Force in Newton's Physics: The Science of Dynamics in the Seventeenth Century*, London: Macdonald; New York: American Elsevier.

Westfall, Richard S.: 1972, 'Newton and the Hermetic tradition', pp. 183–198 of Debus (1972), vol. 2.

Westfall, Richard S.: 1975, 'The role of alchemy in Newton's career', pp. 189–238 of Righini Bonelli and Shea (1975).

Westman, Robert S. and J. E. McGuire: 1977, *Hermeticism and the Scientific Revolution*, Los Angeles: William Andrews Clark Memorial Library, University of California. 1, Robert S. Westman: 'Magical reform and astronomical reform: the Yates thesis reconsidered'; 2, J. E. McGuire: 'Neoplatanism and active principles: Newton and the *Corpus Hermeticum*'.

Whiston, William: 1716, *Sir Isaac Newton's Mathematick Philosophy More Easily Demonstrated ...*, London: printed for J. Senex and W. Taylor. Translated from a Latin edition (Cambridge, 1710; second edition, London, 1726). Reprint (1972), New York, London: Johnson Reprint Corporation, with an introduction by I. B. Cohen.

Whiteside, Derek Thomas: 1962, 'The expanding world of Newtonian research', *History of Science* **1**, 16–29.

Whiteside, Derek Thomas: 1964, 'Isaac Newton: birth of a mathematician', *Notes and Records of the Royal Society of London* **19**, 53–62.

Whiteside, Derek Thomas: 1964*a*, 'Newton's early thoughts on planetary motion: a fresh look', *British Journal for the History of Science* **2**, 117–137.

Whiteside, Derek Thomas: 1966, 'Newton's marvellous year: 1666 and all that', *Notes and Records of the Royal Society of London* **21**, 32–41.

Whiteside, Derek Thomas: 1966*a*, 'Newtonian dynamics', *History of Science* **5**, 104–117.

Whiteside, Derek Thomas: 1970, 'Before the *Principia*: the maturing of Newton's thoughts on dynamical astronomy, 1664–86', *Journal for the History of Astronomy* **1**, 5–19.

Whiteside, Derek Thomas: 1970*a*, 'The mathematical principles underlying Newton's *Principia*', *Journal for the History of Astronomy* **1**, 116–138.

Whiteside, Derek Thomas: 1976, 'Newton's lunar theory: from high hope to disenchantment', *Vistas in Astronomy* **19**, 317–328.

Whiteside, Derek Thomas: *See also* Newton (1967–).

Whitrow, Magda: 1971, *Isis Cumulative Bibliography . . .*, 2 vols., London: Mansell.

Wilson, Curtis: 1968, 'Kepler's derivation of the elliptical path', *Isis* **59**, 5–25.

Wilson, Curtis: 1970, 'From Kepler's laws, so-called, to universal gravitation: empirical factors', *Archive for History of Exact Sciences* **6**, 89–170.

Yates, Frances A.: 1964, *Giordano Bruno and the Hermetic tradition*, London: Routledge and Kegan Paul; Chicago: University of Chicago Press [reprinted 1971].

Yates, Frances A.: 1967, 'The Hermetic tradition in Renaissance science', pp. 255–274 of Charles S. Singleton (ed.): *Art, Science, and History in the Renaissance*, Baltimore: The Johns Hopkins Press.

Yates, Frances A.: 1972, *The Rosicrucian Enlightenment*, London: Routledge & Kegan Paul. Reprint (1975), by Paladin (Granada Publishing), Frogmore, St. Albans, Herts.

Yates, Frances A.: 1973, 'Science in its context', *History of Science* **11**, 286–291. A review of the Pagel Festschift (1972), ed. by A. Debus.

Yates, Frances A.: 1977, 'Did Newton connect his maths and alchemy?', *Times Higher Education Supplement*, 18 March 1977.

D. T. WHITESIDE

NEWTON THE MATHEMATICIAN*

'Newton *and* mathematics', not 'Newton's mathematics' in any very narrow sense, is my theme in this paper. Expect no flash and bang of verbal pyrotechnics. My mood will be one of musing – even maudlin – retrospection, not hard analysis or bold historical hypothesis.

As I look back over the two decades of my life which I have given up to seeking to understand the content and context of Newton's surviving mathematical writings – preliminary worksheets, calculation scraps, finished drafts and printed texts all – I am, I confess, not a little tempted to adjoin to my title the Proustian subhead: 'In search of time lost'. Certainly, I often rub my eyes, wondering at how swiftly time flits by, one year fusing into another, and with so very little accomplished in enduring substance. I have recently been pictured, certainly not without some justice, as happily hewing and chipping away in my mathematical quarry, fashioning and shaping with a craftsman's pride the raw Newtonian stone which I chisel out, well removed from the bustle and grind and whine of the Satanic mill where the heavy industry is located.[1] Just now and then, an outsider peeps briefly in through the entrance to inquire when next I will cart the latest batch of my editorial sculptings out into the outside world for others to share their granite strength and marble beauty. Well, if the cap fits I must of course wear it. But now the time is close at hand to close the quarry down, its stone exhausted, and for me to look up and around from my labour of love. What shall I say in summary of it all?

Not, please, to dread that I shall now just pass singly to list Newton's manifold advances in the diverse areas of technical mathematics of his day, regally reviewing in line abreast his individual achievements in algebra, trigonometry, numerical analysis, synthetic and analytical geometry and infinitesimal calculus, along with their variety of application in mechanics, theoretical optics and planetary astronomy. Rather, I intend something of what A. N. Whitehead sought to elucidate many years ago in a now forgotten Platonic lecture on 'Mathematics and the Good':[2] an overview, sketchy and shadowy in its detail, but broadly illuminating in its import, wherein I may scatter a few of the generalities which have impressed themselves upon my awareness these past twenty years.

Z. Bechler (ed.), Contemporary Newtonian Research, 109–127.

A mathematician is born and he is educated. On the genetic origin of Newton's intellectual genius and its putative socio-economic *cum* psycho-sexual conditioning as he grew up in his Lincolnshire homeland I have very little to say. It seems to me wholly inappropriate to attempt, with C. M. Cox,[3] to assess his intelligence quotient by applying a standard test of so loosely quantifiable an imponderable to the little of his childhood that is known, even through the rosy hue of romanticised hearsay. I do have some slight sympathy for anyone who sees in Isaac a unique vital spark momentarily uniting the earthy Newtons of Woolsthorpe, rising through the generations from the poverty and obscurity of their yeoman-farmer forbears, and the gentle-born Askews of Market Overton, slowly descending the same social scale from their lordly Tudor heyday. I find emptily extravagant when not totally absurd – forgive me this blindspot – the Eriksonian efforts of one recent biographer to conjoin the scientific achievement of Newton's *annus mirabilis* in the middle 1660's with his conjectured childhood 'longing' for a dead father whom he never knew from beyond the womb and for a well-loved mother equally snatched from him, once born, by quick remarriage to a sexually fertile older man, thence positing 'a relationship between this longing and a later intellectual structure in which a sort of an impulse or attraction is a key term descriptive of a force'.[4] Why not go the whole hog and, with R. W. Meyer,[5] look upon Newtonian attraction as a grand Baroque symbol of a political 'power', emanating from a centre somewhere between Court and Parliament, which maintains the locally sovereign bodies, as they go their several ways, in a perpetual if ever-fluid mutual balance? Or choose your own metaphor from Tolstoy, Marx, Spengler or whomever you prefer of the Messiahs of universal history still living.

In my own simple-minded way I must myself go back to the recorded past, however faintly and distortedly it has endured to the present day. First to the dame schools at nearby Skillington and Stoke which taught Newton his reading and writing, if not the third of his three R's; and then six miles northwards from Woolsthorpe to the small single-roomed King's School at Grantham, under whose headmaster Henry Stokes he was soon intellectually to blossom out. One can – citing Ben Jonson's thin menu of 'small Latine & lesse Greeke' and adding to it a sweetening of Hebrew – all too readily disparage the rote learning which was instilled at such country 'grammar' schools.[6] But in Newton's case at least it was better than that. At Grantham, penned in Stokes' own hand it would seem, there survives a little pocket-book,[7] dated 1654, which is filled with 'Notes for the Mathematicks'. Much is as you would expect of such a compendium of rules for boys destined one day in the main to run their father's farms. We have 'Directions for survaying' – how 'To measure incloses or peeces in comon fields by the chaine only' – and in sequel a worked instance 'To turne primes [tenths of poles] into Acres'.[8] But then come sixty-five pages on 'The measureing of Triangles [&] Circles' – elementary plane mensurational geometry, we would say – which are of a different order. 'The manner & way to calculate a table of Sines', one reads, 'by finding ye Sides of . . . regular bodyes inscribed in a Circle whose Radius is given', together with an 'Example of makeing a septuagon' – understand its periphery and area, given its diameter, and *vice versa* – with the accompanying prescript that 'Every Circumference is more then triple his Diameter by such a proportion as is more then $\frac{10}{71}$ & lesse then $\frac{1}{7}$'. Three centuries on, did I at my own North-country grammar school thirty years ago learn how to place a 7-sided equilateral figure within its circumscribing circle? had I heard of Archimedes' lower bound to pi? I think not. And I am impressed that such things were taught at Grantham in the middle 1650's when Newton was his headmaster's prize pupil. Evidence that he himself did imbibe such elements of geometrical drawing may be glimpsed in the interwoven nest of cyclic polygons yet to be seen incised in the plaster of a bedroom window-nook at the Woolsthorpe farmhouse which was, out of termtime, Newton's boyhood home.[9] Years afterwards when he came at Cambridge to study François Viète's mathematical work, his interest was still very much in ruler-and-compasses construction of the regular heptagon and enneagon which are 'trew onely mechanically', and also in determining the side of a 'square equall to y^e area of a circle' in ways built upon the 'Pseudomesolabe' (as Viète dubbed it) $\pi = \frac{3}{5} \times (3 + \sqrt{5})$.[10] When young Isaac went up to Trinity College in 1661, however else he may have stood in diffident awe at the expanses of scholarly learning suddenly opened up to him, in arithmetic and in mensurational geometry he was no beginner.

To Cambridge in 1661. I need scarcely remind that in that first year of restored monarchy its undergraduate scheme of study held little enough to excite the budding mathematician. The notebook[11] in which Newton set down many of his early reading notes confirms that he was ground through a scholastic mill which had changed little for centuries past: a first entrance to a *trivium* of logic, ethics and rhetoric, followed by a *quadrivium* which gave basic instruction in the tenets of Aristotelian physics along with a little Euclidean geometry, all tested in the verbal thesis and antithesis of formal disputation degenerated to be a near-empty stylistic exercise. One could argue that this brought to Newton its long-term benefits in disciplining him to the orderly and forceful presentation of logical argument which he displayed in his mature reasoning. One might a deal more tentatively surmise that his enforced student grounding in Euclid's *Elements* had an equally lasting effect – that its elegant (sometimes not so elegant) deductions of the properties of triangle, quadrilateral, circle and sphere from a few received axioms and postulates led him easily in later years to frame an extended Euclidean space in which uniform kinematic and enforced dynamical motions are analogously defined. But one should be wary of concluding that the timeless niceties of classical pure geometry were ever his *forte*. You all know the story of Newton being put to 'second posing' at Trinity when tested in his 'quadrivium' year on his knowledge of the *Elements*. As he told it to de Moivre in old age,[12] his own account is that when first he met with the *Elements* he 'read only the titles of the propositions, which he found so easy to understand that he wondered how any body would amuse themselves to write any demonstrations of them'; and changed his mind only when he came to the proposition, I,47, 'that in a right angled Triangle the square of the Hypothenuse is equal to the squares of the two other sides'. A story is only a story. But look at Newton's copy[13] of Barrow's 1655 Latin epitome of the *Elements* which was the prescribed student's text, and you will find that in his reading of its first book he was indeed stopped by this theorem of Pythagoras – and to convince himself of its truth he was put to inserting additional construction lines in its figure. That he should have been halted at this basic metrical theorem for Euclidean space is to me significant: it found no place in the little Grantham book of 'Notes for the Mathematicks' – and, of course, if you deny its truth you are into the treacheries of non-Euclidean realms.

Leaf further through the pages of Newton's copy of Barrow's epitome of the *Elements* and you will see more and more the cast of his youthful mathematical mind. In Book 2 he has taken every care to spell out the algebraic equivalents of its linear and quadratic identities, and its geometrical solution of the general quadratic equation by 'completing the square'. On the pages of the ensuing 'geometrical' Books III, IV, VI and XI–XIII – other than to convert to arithmetical equivalent the expressions for the sides of the regular dodecahedron and icosahedron inscribed in a unit-sphere, which Barrow cites *ex Herigonio* in supplement to the last – he has set not a word of comment or query. In contrast, he has fulsomely annotated the intervening 'arithmetical' Books V, VII and X where Euclid presents his Pythagorean-*cum*-Eudoxean notions of real and rational number-ratios and of surds. Were I to tax your patience, I could give you instance upon instance in his later writings[14] where Newton betrays a similar disinterest in classical pure geometry. In the early 1680's and again ten years afterwards he was, it is true, to devote much time and effort to resolving such particular problems of the 'ancients' as that of the Euclidean 'locus to 4 lines' and Apollonius' of 3-circles tangency; more widely – founding his restoration, like Viète, Fermat and others before him, upon Pappus' lemmas on the lost works of Euclid and Apollonius on porisms and other species of *τόποι ἀναλυόμενοι*, 'discovering loci' – he worked hard to recover the analysis underlying the post-synthesised arguments in which the Greeks chose to couch their achievements in higher geometry. But each time he was drawn quickly on, to abstract therefrom univeral canons for invariance under projective linear transformation, and to formulate generalised notions of birational point-correspondence. Pardon the fleeting skip into technicalities. Grant me my point that classical Euclidean geometry played only a minor rôle at best in framing the mould and content of Newton's mature mathematical achievement.

What appears to me ever more clearly to be the decisive event guiding the future direction of Newton's mathematical thought and practice was his reading of Descartes' *Geometrie* in the late summer of 1664. To Newton, as *mutatis mutandis* to Christiaan Huygens more than twenty years earlier when he first read Descartes' *Principia*, the work came as a revelation of the endless bounds of the possible. You will perhaps recall Huygens' own remembered reaction to his Cartesian mentor: 'It seemed, when I read this book, that all was going the best in the world, and I believed when I found some difficulty in it, that it was my fault for not properly grasping his thought'.[15] The *Geometrie* – and to some smaller degree Descartes' *Dioptrique* and *Principia* – must have had the same immediate impact upon Newton. In his case, I regret, we have only a clumsy account, third-hand from his lips by way of de Moivre and Conduitt,[16] telling dully of his crawling progress through the book, reading and re-reading its pages 'till by degrees he made himself master of the whole'. But the thick wad of Newton's research papers surviving from the later months of 1664[17] stand firm witness that it was indeed from the hundred or so pages of the *Geometrie* that his mathematical spirit took fire.

The crucial importance for him of Descartes' short eassy lay not so much in its evocation and illustration of the power of coordinate geometry in resolving problems fearsome to treat by classical methods; though the *Geometrie*'s reduction of the Euclidean '4-line locus' to the standard Apollonian defining 'symptom' of one or other species of conic section showed him the way how analogously to distinguish the general cubic curve into its numerous component species in a manner which he continued to refine and polish over the next thirty years. Nor did its chief significance for him reside in its other novelties of technique; even though its construction of the subnormal at a point of a curve was the foundation of his own generalised insight into the derivative of a 'smoothly' continuous function, as we would now put it. Above all, I would assert, the *Geometrie* gave him his first true vision of the universalising power of the algebraic free variable, of its capacity to generalise the particular and lay bare its inner structure to outward inspection.

To the bare elements of algebra Newton had, a little while before this late summer of 1664, received a beginner's introduction through William Oughtred's *Clavis Mathematicæ* (Key to Mathematics). The third book of the *Geometrie* now carried him on to grasp the niceties of reducing cubic and quartic equations to canonical form, and so to standard solutions which the free variable allowed to be encapsulated in one single formula, once for all. But, as Newton came also quickly to understand, equations – unless these be, as we say, identities – confine the algebraic expressions which they connect, permitting them only limited possibilities of existence. In contradistinction, a true free variable allows what Newton was soon to call a 'universal arithmetic' of quantities and magnitudes, unimpeded in its capabilities of particular representation. And in geometrical counterpart the way was open – as Descartes let his reader but momentarily glimpse – to a similar universal calculus of all that is spatially definable. If you wish, the geometrical plane which for Newton in his youth had been laboriously marked out by ruler and compasses into measured rectilinear and circular plots was now a communal field in which complexly curving lines could be marked out at will, their intrinsic properties mapped out and hierarchically ordered. Gone were static boundaries and their markers: in their place freely variable points whose traces, 'loci', could be specified generally by defining relationships with respect to base-points and base-lines of whose particular positions they are yet essentially independent. With respect to those 'origins', 'axes' and the co-ordinate line-lengths departing therefrom – any of a whole variety of equivalent systems, generalised Cartesian and polar and combinations of these, which Newton's fertile ingenuity soon contrived – the notion of an arbitrary 'geometrical' curve could now be formulated as one representable by a corresponding algebraic defining relationship between them. The road was open to explore the slope and 'crookednesse' (curvature) of such curves in a general way, to frame procedures for evaluating their areas and arc-lengths, and for 'analytically' determining the location of their inflexions, nodes and cusps, for ordering them according to the algebraic degrees of their defining equations: an 'obvious' mode of classification which Descartes had not seen point in adopting, but one whose acceptance affords at once the fruitful criterion of invariance of such degree under geometrical transformation, as Newton was quick to seize upon.

And so on, and so on. In the 'magical' twenty-five months from September 1664 which tradition, following his own preference in old age, allows to be Newton's *annus mirabilis* a mathematician was born. Never did seventeenth-century man build up so great a store of mathematical expertise, much of his own discovery, in so short a time.[18] By freeing the upper bounding ordinate of the series of curvilinear areas through whose interpolation Wallis once hoped to attain the exact quadrature of the circle – but had, in lieu, to settle for his infinite product for $4/\pi$ – Newton was able straightforwardly to derive the series expansion term by term of its general binominal ordinate, and thereafter analogously to give series quadrature of curves whose ordinate can be approximated by any converging polynomial of 'infinite' degree. From Descartes' mode of constructing the subnormal to a curve he passed to determine a universal rule for finding its tangent-slope; and then, in the steps of Heuraet, to demonstrate that this method of tangents and that of quadrature of curves are inverse operations. Make the base variable in any corpus of algebraic expressions one of uniformly flowing 'time', as Newton was to do in autumn 1665 – whether or no he was inspired to do so by attending a contemporary Lucasian lecture of Isaac Barrow's; call the 'speed' of any other fluent quantity with respect to this uniform flow its 'fluxion'; and you have the essence of Newton's analysis by fluents and their fluxional 'moments' of related 'instantaneous' increase or decrement. Add the insight, which he attained by the middle 1680's,[19] that the higher-order fluxions of a fluent quantity are, when divided by the factorials of their order, the successive coefficients of the advancing powers of that fluent in what we call the 'Taylor' expansion of its increment, and the path is wide open to the subtle infinitesimals at which he became so adept in his mature years, and in his *Principia* not least. Descend again to the realm of the numerical, and at once there are waiting to be gathered in offshoot the formulas of finite-difference interpolation and of approximate quadrature in whose concoction and employment Newton was without peer in his day . . . .

I am into technicalities again, and to a self-defeating end. Newton's many-faceted mathematical advances are properly to be savoured in their depth only through long hours of private study, with much sucking of one's pen and creasing of the forehead. To their full and just appreciation I know no royal short-cut. The texts of his extant mathematical papers are now to be found collected within the green jackets of eight bulky volumes,[20] upon whose quality and effectiveness it would be narcissistic for me to comment. Well done or no, distorted in their blinkered editorial vision as might be, you may choose – if not, like Dr Johnson when asked his opinion of women preachers, to compare the performance to that of dogs walking upon their hind-legs – at least to murmur your sympathetic wonder that the thing is done at all.

Be that as it may, do let me insist upon what may appear to you a vacuous truism: that a man who makes and goes on making mathematical advances so profound and so far-reaching as Newton did achieve is – whatever else – a mathematician to his toe-tips. From the first, he regarded everything in his view with an outward eye attuned to arithmetical and geometrical niceties, and with an inner vision which sensed the mathematicisable under-structure in all things. General historians of past scientific development rightly concern themselves, even in their most 'internalist' studies, primarily with mathematics in her applied rôle as bonded servant to mechanical, physical and astronomical advance. In their impatience to be hard at the ontologies and methodologies of scientific theories or head down into the concrete workaday realisms of scientific practice, they so often just brush mathematical considerations aside as mere ancillary technicalities wholly secondary to what is important in exploring the unending past quest for the truth underlying observable phenomena, and for its viable codification in predictive rationale. And yet, is it not true that the book of nature which was for Galileo set out in straight lines and circles – and, let me be fair, an occasional parabola – was not penned for him in more complicated form largely because he had no grasp of geometrical curves of higher order? To Newton, who – put his analytical and projective listings together – knew all 78 species of cubic,[21] and also many a curve of higher degree and of transcendental order, the natural world was writ in a correspondingly complex and intricate form.

To take up, in passing instance, a theme which is well travelled. The field of dynamical motion, in which Newton achieved his most conspicuous success in exact science, had by the 1660's been dug over many times, and was left littered in disarray with heaps of muddy earth and barren, stony soil between which – to continue the simile – there ran only a few beaten paths, and those ending in quagmire. In his youth and early maturity, during the decade from the middle '60's, Newton himself most often took a Cartesian track across which all too soon ended in a boggy vortical swirl.[22] Try as he might to pass through the glutinous morass, he could not succeed, even by conjuring up out of nowhere the chimæra of an inwardly attractive 'æthereall spirit' which (to cite an unpublished sentence of late 1675) 'may not only cause ye gravity wch acts here' – at the earth's surface – 'but ballance ye Moons endeavour to recede from ye central earth she moves about':[23] an æther so thin in substance and credibility that he soon exorcised it in the laudably wise decision to 'avoyd that slippery kind of Philosophising wch consists in guessing at ye figures & motions of ye particles of bodies'.[24] He cut free of the constrictions of such vortically generated, centrifugal Cartesian 'endeavours' only in late 1679 when, as you all know, Robert Hooke wrote out of the blue, urging him to fix his attention solely to 'compounding the celestiall motions of the planetts of a direct' – rectilinear – 'motion by the tangent & an attractive motion towards the central body'.[25] Following that newly signposted track Newton quickly broke through to dynamical enlightenment. By coupling Kepler's law of areas – which he soon realised was valid for orbit in any central-force field – with a differential property of the ellipse which he with some difficulty proved to hold at all its points, he was within weeks able to demonstrate Hooke's desired 'proprietys' of the motions of the solar planets, *viz.* that these be traversed under a 'central attractive power' – directed to their common focus in the sun, that is – 'which makes the velocitys of Descent from the tangent Line or equall' – uniform – 'straight motion at all Distances in a Duplicate proportion to the Distances reciprocally taken'.[26] And thereafter from these two 'axioms' of motion – the Cartesian assumption that all unforced motion be uniform and in a straight line, and the generalised Hookean one that any external force disturbing that inertial path be measurable in magnitude and direction by the deviation away from it which it instantaneously generates – he was to build up most of the nearly hundred propositions 'De motu Corporum' which form the first book of his ensuing *Principia.*[27] 'Tis all mathematics' you must allow as you go through those propositions, one by one, noticing the lavish care which Newton there[28] devotes to such wholly unrealistic central orbits as the Cotesian spirals traced in an inverse-cube field. Mathematician through the ages, from Roger Cotes, de Moivre and Johann Bernoulli[29] onwards, have always been equally fond of this and similar exercises which display only expertise in the exact resolution of differential equations of motion. To others such things will seem only dry and barren theoretical fiction.

Let me go no further here than elsewhere into the technicalities. Although I am, in all my bias and predilection, sorely tempted indeed to plead that no one who has not directly encountered Newton's technical skill and mathematical power, in one context or another, has full right to the name of 'Newton scholar', however experienced he or she may be in the wider subtleties of his scientific and intellectual thought. We must never omit to take into account the over-riding mathematical cast of his mind, even in areas where it does not seem obviously to intrude. To underline the point let me say just a little from my own viewpoint of the *vis insita*, 'innate force', which Newton introduces in his 1680's schemes 'De motu Corporum' continuously to sustain the uniform rectilinear motion of a body in the absence of an external force; and also of the *vis exercita*, 'exercised force', which acts instantaneously in response to an outer force, fractionally changing its size and direction. What is the point, so many have asked in recent years? Why does Newton see the need to invoke such a pair of internal forces at all when elsewhere he shows that he has need only of a unique external diverting force to generate curvilinear, dynamically accelerated motion? We now have a variety of learned expositions[30] tracing the Aristotelian prehistory of *vis insita* in late medieval notions of a body's 'impetus', and delineating the ontological subtleties of both it and its companion *vis exercita*; yet others minutely catalogue the developing sequence and verbal variants of their preliminary statement in early draft and the afterthought of interlineation in Newton's private papers. Which all worthily contributes to a finer, more complete understanding on our part of the source and mould of his developing dynamical thought. But look with a mathematician's eye and see how the problem presents itself. When force conjoins only with force, how may we compound the instantaneously uniform speed of a body's inertial motion with the 'pull' or 'pressure' of any oblique *vis impressa* which acts to quicken or slow it and maybe divert it into a new direction? In some sense, what *more* natural than to resurrect the outworn concept of an 'innate force' sustaining the undeviating uniform motion from instant to instant – but this time as an ideal, a noetic, force which would, *were* it physically real, impel the body in that same instant of time to attain its given speed from rest? *Et voilà*! if you do not blink too hard at positing a *vis innata* which is indefinitely greater than the 'external' diverting force – Newton himself would never have given it a second thought – the rest is simple vector analysis. And the companion *vis exercita* fills an equally natural rôle as the equally ideal force exerted 'internally' on the body as it responds, from instant to instant, to the component of the impressed 'external' force in its direction of motion, bumping up its uniform inertial speed from moment to moment.

It would be easy to draw similar lessons in that other major reach of Newton's prowess in exact science, refractive optics. Out of the welter of corpuscularian and mediumistic theories of the propagation of light through an interface which were expounded in his day, he opted pre-eminently (as I need not say) for a modified Cartesian version of the former. But surely it was not his least reason for doing so that this was amenable to the mathematical techniques he already possessed. Scientific models are ever the art of the formulable and the workable. To determine the limit-intersection of rectilinear rays of light – oh, Euclidean abstraction! – one needs but to argue from first, infinitesimal principles: to compute the changing front of a Huygenian wave as it is transmitted through a medium of varying density requires, other than in the simplest cases, a theory of orthogonal families of curves not to be developed till the early decades of the 18th century. As elsewhere – think again of those near-irrelevant Sections 4 and 5 on conic geometry in the *Principia*'s first book – so in his optics, too, Newton appears often to be more the mathematician at play than scientist in earnest. Why else those long passages in the early pages of his Waste Book where he enthusiastically takes up Descartes' challenge to construct the anaclastic surface which refracts all rays issuing from a point, and broken at a first interface, onwards into a second point – to solve it on a later folio in a form which, but lightly reshaped, he took care to publish in the *Principia* as the final proposition of Book 1?[31] Or the digression in his *Lectiones Opticæ* into the tertiary and quaternary spectra cast by spherical droplets, computing the general formulas which yield their bow radius[32] – but in the case of rainwater stopping short of making the numerical calculation which shows that these encircle the sun? Did Newton's notion that light 'rays' are the tracks of speeding light corpuscles bent out of straight at an interface by a refractive force perpendicular to it have any source or basis more tangible than the distant wham of a Cartesian tennis racquet or his mathematical capacity to define the orbits of such mini-planets of light *quanta* under the instantaneously diverting pull of a *vis refractiva*? We may readily grant that its application to model the curving path of a ray of stellar light through the earth's atmosphere made a pretty instance of his general theory of central-force trajectories; but did it – despite his confident assertion so in his letters to Flamsteed in the autumn and winter of 1694/5 – work other than wholly to mystify the latter?[33] Must he himself not have despaired of the physical reality of the theory when he came in late 1705 to reckon just how many powers of 10 stronger than gravity such a *vis refractiva* must be if the interface be no thinner than a hundred-thousandth of an inch? and at that he carelessly omitted several noughts from his answer, as Bechler has shown.[34]

Not to deny, of course, the frequent times when Newton did evince a very practical side to his mathematical psyche – and particularly so where money was concerned. No man ever took more pains in computing the interest due on a loan, or in determining the present value of an annuity. I always think here of a document of about 1709 to which J. M. Keynes first drew notice nearly forty years ago[35] in presenting it to Newton's *alma mater*. There, in response to a request from Richard Bentley to intercede between him and the Fellows at Trinity over how the Master's share of college income should be reapportioned, Newton employed an approximating parabola to compute his basic stipend from his livery allowance under the old statutes, and thence deduced a fair linear apportioning of Master's to Doctors' to M.A.'s salary – but taking care to insert a low proportionality factor, and then adjoining a lump sum of £400 to cover Bentley's expenses in public duty and social entertainment. (Not, I might add, that Bentley would have anything to do with Newton's scheme, since it would have lost him more than 10% of the sum for which he successfully stuck out. So must perish all best-laid plans of fair-dos for all. . . .) I could cite not a few similar computations of Newton's in astronomy – calculating, for instance, the apparent path of the 'great' comet of 1680, or averaging Flamsteed's submitted sightings of the moon to find a mean value for a lunar inequality. This is workaday mathematics in her rôle as subservient handmaiden, one which I need not further elaborate upon – though ever to be on guard against those occasions when Newton makes her mistress in the delicate art of numerical cookery.[36]

Never ever to forget that mathematics had for Newton, above and beyond its place as trusty tool-kit, an inner beauty and strength independent of all outward motivation and application. To those who are insensitive to the elegance and power of mathematics as an intellectual discipline in its own right, shorn of all footholds into the world at large, it is difficult to put over just why Newton should have interested himself (as he did at various times in his life) in such 'useless' topics as projective and transformational geometry or Diophantine number theory or the exact quadrature of algebraic multinomials;[37] why it was important for him to take forty pages of his *Principia*[38] to give solution of problems of circle-tangency and of anharmonic conic theory; why it mattered to him to seek numbers which can be expressed in two separate ways as the sum of two cubes[39] (he, unlike G. H. Hardy two centuries later, would have needed no prompting from Ramanujan to appreciate the significance of 1729 in this respect); why he should have spent so much effort in determining the most general Cartesian equation of the quartic trefoil from the condition that this be invariant under rotations through 120° about its centre, cheerfully setting himself to reduce the 106 terms of the ensuing transformed equation;[40] why he just *had* to go on deriving formula after formula, each more general than the last, for the squaring of curves by series 'wch brake off'.[41] There you have a 'pure' mathematician, as the old phrase has it, at times wholly absorbed in his Cambridge ivory tower in elaborating theorems and properties and algorithms and elegances of construction for their own sake; and magnificently did he practise his talents and expertise. There was no more gifted, no more widely versed mathematician in all the world in his day; no one more adept in algebra, more skilled in geometry, more adroitly wise in the subtleties of infinitesimal variation. Set him against James Gregory, Leibniz, Huygens, the Bernoullis Jakob and Johann, and in his heyday he was their peer in the mathematical realms in which each of these showed themselves at their best. Lagrange's aphorism that Newton was lucky to be born at just the right time to be able to unlock the secrets of the universe is not without truth; and there were giants a-plenty around with shoulders broad enough to raise him on high to the keyhole. But do give full measure to the trained acuity, the deep-read learning coupled with originality of approach, and – let me say it – the fund of genius which allowed him to insert the key and fully turn it in his *Principia*, the *Philosophiæ Naturalis Principia Mathematica*.

In its manifestation, that genius was never the seemingly quicksilver thought and deft Pascalian insight of so many lesser-gifted men, but a deeper, firmer, more solid and enduring sort. As you all know, Newton himself in 1692 epitomised his intellectual achievements as 'due to nothing but industry & a patient thought':[42] which is not to say that his mental stamina and stubborn sweat was but average and commonplace. The power of intense, unremitting concentration which he developed at Cambridge in the 1660's – by all account he could work a 16-hour day, 7 days a week for months on end – was to remain with him undimmed over the next quarter of a century, and was gradually to be underpinned by a massive weight of knowledge and technical expertise, in mathematics as in the other areas of his scientific studies. And behind the concentration – do I show my bias in thinking so? – was an earthy core of straight-thinking common sense. In all the thousands of sheets of his preserved mathematical papers I can detect no hint of any belief by him in number-mysticism, no trace of the extravagance of hermetic mathesis: no neo-Pythagorean arithmologies lurk there that I know of, no cabbalistic *gematria*, no cryptic magic squares or motifs in John Deeist style complexly intertwining pentacles and stellate heptagons. Elsewhere it may be so, but I need to be shown. Karin Figala has just lately directed our overdue attention to the three simplest non-trivial solutions of the recursion $a_{n+1} + 2a_{n-1} = 3a_n$ which Newton set to be basic in his schemes for determining the successive degrees of rarefaction in 'sulphurial' and 'mercurial' and other substances, adding in comment that 'In this way' – combining one step forward and two backward! – 'generation of different degrees are connected in a "trinity" '.[43] Maybe so in that shadowy no-man's-land between renaissance alchemy and modern chemistry. But had any one else put such an interpretation to him, Newton in his trenchant way would have despatched the metaphorical spirit with the same curt 'See the mystery!' with which he afterwards in a couple of lines[44] surgically dismembered Leibniz' claim to 'harmonic' perfection in summing to a finite aggregate the infinite series of reciprocals of triangular and higher polygonal numbers (binomial coefficients to us). There is the no-nonsense, down-to-earth Newton as I have always met with him.

The iron discipline of his working mathematical practice showed no crack till he came towards the middle 1690's – here murmur of the 'black year' of 1693 if you must, but I can see evidence of only a gradual deterioration in his rock-hard stamina. In April 1695, then past his 52nd birthday, he admitted in confidence to Flamsteed that if 'I set myself wholy to calculations . . . I can endure them & go through them well enough. But when I am about other things . . . I can neither fix to them wth patience nor do them wthout errors'.[45] A year later, foreseeing (as I understand it) the accelerating decline of his creativity with the onset of old age, he abandoned the life of a university research scholar for ever, settling for worldly fame and public office in the metropolis. A brief flash in January 1697 of the erstwhile kingly lion's claw in solving Johann Bernoulli's challenge to identify the brachistochrone curve – a couple of years earlier his method of 'maxima & minima in infinitesimals' would have detected that this is the cycloid in a few minutes, not the twelve hours he in his rustiness then took[46] – and his years of mathematical supremacy were at an end. Ever after, as he phrased it to Flamsteed in a notorious outburst two years later, – the immediate context was his largely abortive 'New Theory of the Moon's Motion', but read in its stead all exact science – he was 'concerned to be publickly brought upon ye stage about what perhaps will never be fitted for the publick'.[47] When in February 1716, his 73rd birthday past, he tried again to respond to a new challenge problem from Bernoulli's pen – this time to determine orthogonals to families of curves whose parameter is absorbed into their defining differential equation – all was purblind feebleness.[48] Newton's thin essay at a solution – as before, published anonymously in the *Philosophical Transactions* – this time so far failed to elicit from Bernoulli any recognition of a leonine hand that he confided to Leibniz his guess that Brook Taylor was 'The anonymous Englishman' who had 'not attained the solution even by a finger-tip'.[49] Take pity on the editor of the jejune, myopic products from Newton's mathematical pen in his flagging old age. The death of talent is dispiriting to the onlooker, but the heart truly sags when one has to be close attendant upon the mortification of genius.

Enough. I have said my piece twice over. In his lecture on 'Mathematics and the Good' Whitehead delivered a *dictum* which has stayed with me during the nearly thirty years since I first read him as a boy at school: 'Mathematics', he states, 'is the study of pattern in abstraction from the particular'.[50] Newton learnt that lesson well in the 1660's, and was ever after able most wonderfully and fruitfully to apply it till old age caught up with him. Let me leave you with a misquotation from *Hamlet*, to urge upon you one last time this evening that in his case it really is that

> '. . . mathematics' the thing
> Wherein to catch the science of the king'.

And so to say, 'Goodnight, sweet Prince' – 'Our play is done'.

NOTES

* Annotated script of a lecture delivered on the evening of 8 August 1977 to the participants in the commemorative Newton symposium at Churchill College, Cambridge.

[1] See R. S. Westfall, 'The changing world of the Newtonian industry' (*Journal of the History of Ideas* 37 (1976), 175–84), pp. 175–7; and compare his review of the fifth volume of my edition of *The Mathematical Papers of Isaac Newton* in *American Scientist* 61 (1973), 762.

[2] Printed in his collected *Essays in Science and Philosophy* (London, 1948), pp. 75–86.

[3] In conglomerately studying 'The Early Mental Traits of Three Hundred Geniuses' (in Lewis M. Terman (ed.), *Genetic Studies of Genius*, Stanford, 1926).

[4] F. E. Manuel, *A Portrait of Isaac Newton* (Harvard, 1968), p. 83.

[5] See his *Leibnitz and the Seventeenth-Century Revolution* (Cambridge, 1951), pp. 58–9.

[6] The best general account of what was taught in England's grammar schools in the seventeenth century is still, to my mind, Foster Watson's *The Beginnings of the Teaching of Modern Subjects in England* (London, 1909); his Chapters VIII–X (pp. 254–393) comprehensively treat of the teaching of mathematics (arithmetic, geometry and – this all but wholly in the universities – astronomy).

[7] Grantham Museum D/N 2267 (Hill 31).

[8] In this metric age I may perhaps remind the unwary that a pole is a square rod, that is, $(5\frac{1}{2})^2 = 30\frac{1}{4}$ square yards, whence 1 acre (= 4840 square yards) = 160 poles.

[9] Compare H. W. Robinson, 'Note on some recently discovered geometrical drawings in the stonework of Woolsthorpe Manor House', *Notes and Records of the Royal Society of London* 5 (1947/8), 35–6.

[10] See Newton, *Mathematical Papers* 1 (Cambridge, 1967), pp. 85–6.

[11] U.L.C. Add. 3996; compare R. S. Westfall, 'The foundations of Newton's philosophy of nature', *British Journal for History of Science* 1 (1962), 171–82.

[12] The original of John Conduitt's 'Memorandum relating to Sr Isaac Newton given me by Mr Abraham Demoivre in Novr 1727' is now in the University of Chicago. Its portion

dealing with Newton's mathematical baptism is printed (from H. R. Luard's 1870's copy in U.L.C. Add. 4007, ff. 706^{r}–707^{r}) in my brief study of 'Isaac Newton: birth of a mathematician' (*Notes and Records of the Royal Society* **19** (1964), 53–62), pp. 58–9; and again in Newton, *Mathematical Papers* **1**, pp. 5–6.

[13] Now Trinity College, Cambridge, NQ.16.201[1] (= Item No. 581 in J. R. Harrison, *The Library of Isaac Newton* (Cambridge, 1978); see also its p. 19 and p. 51). On this copiously annotated student text of Newton's see further H. Zeitlinger, 'A Newton Bibliography' (= (ed.) W. J. Greenstreet, *Isaac Newton: 1642–1727* (London, 1927), 148–70), pp. 169–70 (+ 168); and *Mathematical Papers* **1**, p. 12 note 28.

[14] See especially *Mathematical Papers* **4**, Part 2 and **7**, Part 2: *passim.*

[15] In his original French of this reflection on reading Baillet's 1691 *Vie de Monsieur Descartes*: 'Il semblait lorsque je lus ce livre des *Principes* pour la premiére fois que tout allait le mieux du monde, et je croyais quand j'y trouvais quelque difficulté, que c'était ma faute de ne pas bien comprendre sa pensée' (*Œuvres complètes de Christiaan Huygens* **10**, 1905, p. 403).

[16] See note 12 above.

[17] Those reproduced in *Mathematical Papers* **1**, pp. 155–297. I need not here give any detailed markers to the crowd of particular statements and assertions which follow.

[18] The extant record of this youthful giant's leap in mathematics is gathered in the pages of *Mathematical Papers* **1**: *passim*. Again I see no need to give particular supporting references.

[19] See Proposition X of Book 2 of Newton's *Principia* ($_1$ 1687): this is fully analysed in my newly printed Volume 8 of his *Mathematical Papers*, where (Part 1, 6) the recastings of its argument which Johann Bernoulli led him to make in autumn 1712 are presented in their essential totality. The general 'Taylor' expansion of an incremented *quantitas* is found explicitly stated by Newton only in the first (1691) version of his treatise *De quadratura Curvarum* (first published in *Mathematical Papers* **7**, 1976, Part 1, see especially pp. 96–8).

[20] *The Mathematical Papers of Isaac Newton*, 8 vols., Cambridge, 1967–81. A short volume of indexes will complete the edition.

[21] See *Mathematical Papers* **7**, pp. 194, 565.

[22] Most notably in the elaborate theory of the moon's motion, as I pass to insist. In the end-papers of his library copy of Vincent Wing's *Astronomia Britannica* (London, 1669) – this is now in Trinity College, Cambridge (shelfmarked NQ.18.36) – Newton entered, around 1670, a complicated theory of solar, terrestrial and lunar vortical interaction to 'explain' the principal inequalities in the weaving orbit of the moon; compare my overview of 'Newton's lunar theory: from high hope to disenchantment' (*Vistas in Astronomy* **19**, 1976, 317–28), p. 318, and also my earlier survey of 'Newton's early thoughts on planetary motion: a fresh look' (*British Journal for History of Science* **2** (1964), 117–37), p. 127. It is regrettable that Newton's vortical theory itself remains unpublished.

[23] See U.L.C. Add. 3970.3, f. 536^{v}. It is fairer to the subtleties of Newton's supposition about the properties of the 'æthereall Medium much of the same constitution with air, but far rarer . . . ' which he postulated to transmit light in his 1675 'Hypothesis' (*vide Correspondence of Newton*, **1**, 1959, pp. 364 ff.) to cite this unpublished conjecture in full, *viz*. that 'as this [æthereall] spirit descends [to ye earth] it may not only cause ye gravity wch acts here but ballance ye Moons endeavour to recede from ye central earth

she moves about, & [also affect] ye Lunar system by straitening a little ye course of y^{e} æther between her & ye earth & so pressing it towards ye earth & thereby making ye earth [recede] ye contrary way. And besides this æthereal sp^{t} ye main body of [flegmatic] æther may cause divers effects & motions as ye carrying ye Planets about in their orbits. . . . For if y^{e} æther about ye earth move in an Elliptical vortex & be comprest by ye Lunar system after ye manner described by Des-Cartes, y^{e} æther about ye superficies of ye earth must tow[a]rds y^{e} [Moon] & on ye other side of ye earth opposite to her descend a little by compression & in ye other quarters ascend as much by relaxation, & by those motions (not by mere pressure) increase ye gravity of all things a very little . . . '. (The last notion is set out more elaborately in Newton's 1670 annotation on Wing's *Astronomia*; on which see the previous note.) Two pages earlier (f. 535^{v}) in discussing the 'secret [*sc.* undetected] principle of unsociableness' which 'lingers' in the physical world – allowing, for instance, 'oyle & water . . . freedom enough to mix wth one another yet . . . keep asunder' – Newton had guessed that perhaps 'the like unsociableness may be in æthereal natures' with 'ye Planets' having 'vortices of æther unsociable to y^{t} of ye sun's'.

24 The directness of this unusually honest phrase (a late insertion in a related context on U.L.C. Add. 3970.3, f. 474^{v}) was, of course, too outspoken for Newton to pass it on even in the semi-privacy of a letter to Oldenburg, and it is here rescued from oblivion.

25 Hooke to Newton, 24 November 1679 (*Correspondence* 2, 1960, p. 297).

26 As Hooke specified in his following letter of 17 January 1679/80 (*Correspondence* 2, p. 313). I have sketched how Newton might have gone on to achieve the crucial insight that Kepler's law of areas is basic to analysing all central-force orbits (see my 'Newton's early thoughts on planetary motion . . . ' (note 22 above), pp. 135–6), but no documentary evidence survives which might show whether I am nearer to the historical truth than the others – notably J. W. Herivel and R. S. Westfall in recent years – who have given differing reconstructions of the sequence of Newton's thought at this time.

27 *Principia* (London, $_1$1687), pp. 1–235. The several preceding states of this 'De motu Corporum Liber primus, as they incompletely survive, are reproduced in *Mathematical Papers* 6, Part 1, along with my editorial appreciation of their mathematical significances.

28 *Principia*, Liber 1, Prop. XLI, Corol. 3; see also Newton's private recomputation of the correctness of his stated construction of the inverse-cube spiral orbit for David Gregory in May 1694 (reproduced in *Mathematical Papers* 6, pp. 435–8).

29 Compare *Mathematical Papers* 6, pp. 349–50, note 209 and 353, note 213.

30 By I. B. Cohen, A. Gabbey, J. E. McGuire, E. McMullin, N. Tajima and R. S. Westfall, to cite (in alphabetical sequence) the names but of those who here come immediately to mind.

31 See *Mathematical Papers* 1, pp. 558–9, note 25; and 3, pp. 528–30.

32 *Lectiones Opticae* (London, 1729), Part 1, Section 4, Scholium to Prop. 36 (reproduced in *Mathematical Papers* 3, p. 506 from the autograph original, now U.L.C. Dd. 9.67, which Newton himself titled 'Optica').

33 I treat this matter at length in a lecture – given initially at Aarhus in August 1976, and in revised form at Lancaster in May 1978 – on 'Newton and Flamsteed on atmospheric refraction: the mathematical and the practical' (= *Centaurus* 24, 1980, 288–315). The mathematical theory adduced by Newton derives from Proposition XLI of Book 1 of his *Principia* (see *Mathematical Papers* 6, pp. 350–1, note 209 and especially pp. 431–4).

[34] In his penetrating discussion of 'Newton's law of forces which are inversely as the mass . . . ' (*Centaurus* 18, 1974, 184–222), pp. 189 and 202 (+ 220, notes 7 and 8).

[35] In his letter to Lord Adrian (then Master of Trinity College) on 22 April 1939 (reproduced by A.N.L. Munby as 'A mathematical analysis by Newton of a problem in college administration', *Isis* 49 (1958), 174–6).

[36] R. S. Westfall has percipiently studied three classical instances of Newton's skill in numerical confectionery in his study of 'Newton and the fudge factor' (*Science* 179 (1973), 751–8); see also his *Force in Newton's physics* (London, 1971), pp. 497–50. A further instance of skilled mathematical cookery – that of obtaining the observed mean advance of lunar apsides as a 'consequence' of his best efforts at a solution of the restricted 3-body problem, one to be given proper, accurate solution only by Clairaut in the 1740's – is set out in *Mathematical Papers* 6, pp. 508–36.

[37] See *Mathematical Papers* 4, pp. 304–14; 1, pp. 155–212; 4, pp. 74–108; and 3, pp. 373–85 respectively.

[38] *Principia*, ${}_1$1687, pp. 61–103 [= Book 1, Sections 4/5]; compare *Mathematical Papers* 6, pp. 230–98.

[39] This problem he took, along with the basic algorithm which yields its familiar standard parametrisation, from François Viète's *Zetetica*; see *Mathematical Papers* 4, pp. 88 and 96. For those unfamiliar with the following allusion let me stray to make a brief quotation from Hardy's obituary notice of Ramanujan in the *Proceedings of the London Mathematical Society* (2) 19 (1921), xl-lviii (= *Collected Papers of Srinivasa Ramanujan* (Cambridge, 1927), pp. xi–xxxvi): 'I remember once going to see him when he was lying ill at Putney. I had ridden in tax-cab No. 1729, and remarked that the number (7.13.19) seemed to me rather a dull one, and that I hoped it was not an unfavourable omen. "No", he replied, "it is a very interesting number; it is the smallest number expressible as the sum of two cubes in two different way" . . .' (*Collected Papers*, p. xxxv). Whether it be an ill omen or no, the fact that 1729 splits into the pairs of cubes $9^3 + 10^3 = 1^3 + 12^3$ was familiar to all workers in the theory of numbers from the time of Diophantus up to Euler's day, and only lost to 'common' mathematical knowledge in our modern age.

[40] *Mathematical Papers* 7, pp. 665–70.

[41] In the drafts for the original (1691) version of his treatise 'De quadratura Curvarum'; compare *Mathematical Papers* 7, pp. 140–6.

[42] Newton to Bentley, 10 December 1692 (= *Correspondence of Newton* 3, p. 233).

[43] See her essay – stimulated by B. J. T. Dobbs' recent book on *The Foundations of Newton's Alchemy* (Cambridge Press, 1975) – on 'Newton as alchemist' (*History of Science* 15, 1977, 102–37), pp. 127–8 (and compare p. 137, note 55). When put in the form $a_{n+1} - a_n = 2(a_n - a_{n-1}) = \ldots = 2^{n-1}(a_1 - a_0)$, whence similarly $a_n - a_{n-1} = 2^{n-2}(a_1 - a_0), a_{n-1} - a_{n-2} = 2^{n-3}(a_1 - a_0), \ldots$, by summing these there is evidently $a_{n+1} - a_0 = (2^{n-1} + 2^{n-2} + \ldots + 1)(a_1 - a_0) = (2^n - 1)(a_1 - a_0)$, and so $a_{n+1} = 2^n(a_1 - a_0) - (a_1 - 2a_0)$. Newton's trio of cases are $a_1 = 1, a_0 = 0; a_1 = 2, a_0 = 1$; and $a_1 = -1, a_0 = 0$.

[44] See his anonymous review of his own equally anonymously edited *Commercium Epistolicum D. Johannis Collins et aliorum de Analysi promota* (London, 1712) in *Philosophical Transactions* 29, No. 342 (January/February 1714/15), pp. 173–224; the quotation is from p. 183. (Newton's original draft of this late *addendum* to his '*Recensio*' exists on a paper scrap loose in U.L.C. Add. 3958.3.)

45 Newton to Flamsteed, 22 April 1695 (= *Correspondence* 4, p. 106).
46 I make this point at length in *Mathematical Papers* 8, pp. 8–9.
47 Newton to Flamsteed, 6 January 1698/9 (= *Correspondence* 4, p. 296).
48 Compare *Correspondence* 7, 1976, pp. 290–2. Newton's inconsequential efforts at solution are now printed in full – with pertinent editorial comment from my pen – in *Mathematical Papers* 8, pp. 425–41.
49 'Anglus anonymus (quem Taylorum esse credo) [solutionem] ne apice quidem digiti attigit', Bernoulli wrote in his last letter to Leibniz on 11 November 1716, a few days before the latter's death; see C. I. Gerhardt (ed.), *Leibnizens Mathematische Schriften* 3.2 (Halle, 1856), p. 972.
50 *Essays in Science and Philosophy* (see note 2 above), p. 85.

RICHARD S. WESTFALL

NEWTON'S THEOLOGICAL MANUSCRIPTS

The past two decades have witnessed a concentrated exploration of Newton's manuscripts. More than anything else, this factor has set the tone of recent Newtonian scholarship. However, one major section of his papers remains almost totally unknown. The great bulk of Newton's theological papers have become available only recently. Lost from public view after the auction in 1936, they have come to rest at last as the Yahuda Papers in the Jewish National and University Library in Jerusalem. Professor Manuel used them in the preparation of his recent *Religion of Isaac Newton* (Oxford, 1974), but with his attention focused on Newton rather than the manuscripts, Professor Manuel did not attempt to offer a guide to their content. Such a guide, not confined to the Yahuda Papers but inevitably concentrating heavily upon them since they are the major collection, is what I hope to present in this paper. I have examined all of the theological papers the location of which is known except the one in the Bodmer Library (Geneva), which apparently regards it as a perishable resource which reading would exhaust. Judging by the papers that accompanied it at the sale, I think the Bodmer manuscript contains only miscellaneous fragments without anything not found elsewhere, but only a chance to see it, a privilege which the Bodmer Library does not grant to mere scholars concerned to study Newton, can determine its content. While reading will not exhaust the manuscript, one reading is likely to exhaust any interest in it if my guess at its contents is correct. I shall be able to contain my disappointment on learning that such is the case. Meanwhile, drawing on the other papers, I shall offer a chronological survey of Newton's activity as a theologian, relating it to his career and to themes of his scientific thought.

Despite apparent early concern with religion, manifested in such things as the theological library he inherited from his step-father, the books he is known to have purchased upon coming to Cambridge, and the shorthand examination of his conscience in 1662, Newton's theological study, as his surviving papers reveal it, began only in the early 70's. He was at that time facing the necessity of ordination by the year 1675 in order to retain his fellowship at Trinity; perhaps that need set him on sustained theological reading. Whatever the cause, the reading notes, which I date primarily by the

Z. Bechler (ed.), Contemporary Newtonian Research, 129–143.

hand in which they were written, establish that he did begin at that time. Never one to do things half way, he plunged into an extensive program of study which conducted him through all of the early fathers of the church. The notes from that reading survive in his theological notebook (King's College, Cambridge; *Keynes MS. 2*) and various papers among the Yahuda MSS. (*Yahuda MSS. 2.5b*, ff. 40–7; *5.2*; *5.3*; *13.1*; *14* passim; also *Keynes MS. 4*). If the program of study aimed to prepare Newton for ordination, its most immediate result made ordination impossible. Very quickly, Newton read himself into advanced heresy. Already in the theological notebook, one can find the elements of an Arian Christology. Though Christ was the only begotten son of God, and hence never merely a man, he was not equal to God, not even after God exalted him to sit on his right hand as a reward for his obedience unto death. Among the Yahuda Papers is a sheet from the period 1672–75 with twelve propositions on Christ (*Yahuda MS. 14*, f. 25; *cf.* ff. 173–3^{v} and *2.5b*, ff. 40–7. *Babson MS. 436*.) The propositions are wholly Arian in their outlook. Thus at the very beginning of his theological career, Newton adopted the position which would remain unaltered through the rest of his life and which continued to find expression among the last papers he composed before his death.

Newton's plunge into heresy nearly blighted his career. Late in the 18th century, a story surfaced from the joint recollection of the Uvedale family of a competition between Robert Uvedale and Newton in 1673 for the fellowship in law, one of the two fellowships in Trinity exempted from the requirement of ordination. Resting as it did on nothing more substantial than family tradition passed down through three generations, the Uvedale story is not very secure. Nevertheless, it is believable. Uvedale did in fact obtain the fellowship, by virtue of his greater seniority if we accept the story, as we can since Trinity decided all questions of that sort by seniority. Moreover, Newton did in fact refuse ordination; and in the spring of 1675, as a letter to Oldenburg shows, prepared to vacate his fellowship (*Corres* 7, 387). As we know, he was saved at the last moment by a dispensation from the Crown. Meanwhile, his early conversion to Arianism was the fundamental fact of his theological development. We shall be returning to it repeatedly. For the moment, note only that it may have played a role also in his natural philosophy. To the discussion now in progress on the philosophical source of Newton's conception of God, add the possible influence of Arianism, though I cannot myself see any way to determine which came first, the theological or the philosophical stance. The God of the General Scholium, the Pantocrator supremely dominant over his creation, was not unlike the God that Newton met, and recorded in his theological notes, among Arian theologians. (See *Keynes MS. 4*.)

From the very first, part of his theological study was the interpretation of the *Book of Revelation*. (*Yahuda MSS. 1, 2.1, 10.3*, ff. 14–29.) Newton always considered it to be the fundamental book in the Bible, to the interpretation of which all the rest of Scripture was subordinate. The surviving manuscripts, loaded with heavy revisions, testify that his work on the *Apocalypse* continued on through the 70's. (*Yahuda MSS. 10.3*, ff. 1–12; *14*; *28*; *39*; *Keynes MS. 1*.) One of the persistent problems of Newtonian scholarship is the twenty year delay between his early insight into the inverse square relation and the final publication of the *Principia*. His manuscripts suggest that the primary reason for the delay was the simple fact that through the 70's and early 80's Newton devoted almost no concern to the questions of mechanics, celestial or terrestrial. As we know, he invested extensive time during this period in alchemical study and experimentation. The additional theological study that I have indicated can only be called prodigious. To the interpretation of the prophecies he devoted, it appears, even more time. Looking at his manuscripts, I find it hard to comprehend how he found time for it all; only a Newton could have done so. To have fitted in active concern with the problems of dynamics would have been impossible, or close to it, even for Newton.

Newton's study of the prophecies has been well known from at least the time of his death, when his *Observations upon the Prophecies* were published, though the period at which he began it has not been known. Scholars have generally assumed that this occupation was a recreation of his old age, as indeed the published work was. For reasons about which I can only speculate, he decided systematically to obscure the point of his interpretation, so that I am prepared to defy anyone who has not read the early manuscripts to find any point at all in the formless and aimless chronologies that constitute the published work. There is no similar problem with his early treatise on which, with all the boldness of youth, he emblazoned his point in letters of fire. His point was an adaptation of the central theme of Puritan interpretations with which he was familiar. The "Great Apostacy," which was of course Roman Catholicism, had furnished the core of Puritan interpretations in the writings of men such as Joseph Mede and Henry More, whose influence Newton acknowledged. The "Great Apostacy" also furnished the core of Newton's interpretation, which, in his early writings, unlike the later, published treatise, based itself on the *Book of Revelation* to the almost total exclusion of *Daniel*. With him, however, the apostacy was not Catholicism but trinitarianism. The seven trumpets of the *Book of Revelations* (or better, the first six) and the plagues associated with them in his view, were punishments poured out by the wrath of the Lord on a people who had gone whoring after false gods. The work exhibits a certain tension. On the one hand, Newton adopted a chronology which he found within the *Apocalypse*. He insisted that references to 1260 days, forty-two months, and a time times and half a time (taken to mean three and a half years) all pertained to the same period of 1260 prophetic days, that is, 1260 years, during which the Great Apostacy stood at its height. The early treatise was explicit in setting the beginning of the period, which coincided with the sounding of the fifth trumpet, in 607. By simple addition, it follows that he believed the second coming, at the sounding of the seventh trumpet, to lie two hundred years in the future. In tension with the chronology was a palpable sense of expectation. The meaning of the prophecy was finally being revealed, and with its revelation the end must be at hand. The manifest passion with which he invested the treatise spoke to the same expectation. "Idolators, Blasphemers, & Spiritual fornicators," he thundered at the Athanasians in the silence of his study. And again, borrowing the language of Scripture to add gravity to his curse, "Seducers waxing worse & worse, deceiving & being deceived – such as will not endure sound doctrine but after their own lusts heap to themselves teachers, having itching ears & turning away their ears from the truth to fables." (*Yahuda MSS. 1.4*, ff. 67–8; *1.2*, f. 26^{v}.) One cannot read the early treatise without becoming convinced that Newton saw himself as part of the remnant of the woman's seed persecuted by the dragon, as one of the saints on whom the beast made war, and the tone of the treatise implies the expectation of early release. The battle of the great day was at hand. Against whom was this passion directed? The Roman Catholic Church, which Newton never ceased to hate and fear, stood always in the background. Nevertheless, the specific form of his denunciations of the hypocrisy and greed of the trinitarian church suggests a target closer to home. The denunciations seem to describe what we know from other sources of the Cambridge around Newton. We know also that he was largely isolated from that Cambridge, indeed alienation from it. Newton's study of the *Apocalypse*, or rather his conversion to Arianism, which animated the study, and his consequent decision not to accept the ecclesiastical career he had sworn to pursue at his election to a fellowship, appears to have been first of all a rebellion from his academic surroundings.

Meanwhile, the interpretation of the prophecies embodied a program of study, revealed in the remains it left, of almost incredible dimensions. I say "almost incredible dimensions" since anyone engaged in the study of Newton rapidly learns that what would be incredible for ordinary mortals was possible for him. It was in fact two programs of study. Newton found it necessary, on the one hand, to establish the true text of the *Revelation*. To this end, he returned to the original incarnation of the book, collating more than twenty-five Greek versions both printed and manuscript. To them he added citations of the *Apocalypse* in the early fathers. (*Yahuda MSS. 4.1* and *4.2. Cf. 2.5c*, ff. 1–2 and *7.2j*, ff. 91–4.) So also, he combed the whole Bible, which was meant to contribute to the interpretation of the prophecies, to find corroborating testimony – the foundation of that unequaled knowledge of Scripture to which John Locke testified. The study of Scripture by itself could not suffice, however, for the truth of the prophecies could be established and the glory of God enhanced only by the demonstration of an exact correlation between prophecy and historical event. Although he read the best contemporary historians, Newton was not one to rest satisfied with secondary accounts. Return to the original sources he must in a program of intense historical study which concentrated on the first six centuries of the Christian era, and especially on the fourth century, when the Great Apostacy triumphed. Consider for example his discussion of the silence of half an hour (a period of seven and a half years according to his chronological principles) before the sounding of the first trumpet. Historically, the period ended with the war of Theodosius and Maximus. To date it exactly, Newton followed the movements of Maximus in the early 380's. To this end he consulted the letters of Jerome, the writings of Zosimus and Pacatus, the testimony of Sulpitius Alexander (an eye witness quoted by Gregory of Tours), the letters of Ambrose, the *Annalium Boiorum*, and Gothofredus' commentary (from the early 17th century) on the Theodosian Code. (*Yahuda MS. 1.4*. ff. 55–6). A brief passage of about ten pages concerned to establish that peace broke out for a short time in the year 380 cited Zosimus, Theodoret, Cedrenus, Baronio (a 16th century scholar), Marcellinus, Ammianus Marcellinus, Socrates (the historian), Sozomen, Prudentius, Symmachus, Paulinus, Gregory Nazianzen (two works), Jornandes, Sigonius, Prosper, Joseph Mede, Rufinus, Petrus Diaconus, Platina, Idatius, Victor, Gregory the Presbyter, Claudianus, Orosius, Cassiodorus, Pacatus, Jerome's addition to Eusebius' History of the Church and two of his letters, the Alexandrian Chronicle, the Theodosian Code, and Gothofredus' commentary on the code. (*Yahuda MS. 1.4*, ff. 33–42.)

The interpretation announced itself as the work of the man who had recently composed *De methodis serierum et fluxionum*, the definitive early exposition of his fluxional method, and would soon compose the *Principia*. He set himself the goal of methodizing the study of the prophecies. Thus he began with fifteen rules of interpretation (*e.g.*, assign only one meaning to one figure, prefer the simplest and most literal meaning). He proceeded on to a catalogue of seventy figures or types most used in the prophecies, devoting some care to achieving exactly the number seventy. Next came a section entitled "The Proof," which searched the whole Bible and other relevant ancient sources, such as the Arabian Achmet and the Helenistic Greek interpreter of dreams, Artemidorus, to establish the meaning of the figures from other examples of their usage. Newton's goal, which he stated explicitly, was to reduce the interpretation of the prophecies to demonstration free from the vagaries of individual fancy. (*Yahuda MSS. 1.1* and *2.1*.) Passages in the early sections of the treatise sound like the Fourth Rule of Reasoning in Philosophy which he added to the third edition of the *Principia* or, what is probably more relevant, methodological passages in his letters on optics in the early 70's, at much the time when he wrote the early interpretation. Initially, he also laid out the treatise in "Propositions," a title he changed in revision to mere "Positions."

In the late 70's Newton embodied much of the same material in a history of the early church. (*Yahuda MSS. 2.2, 2.3, 2.5b, 5.3, 11, 12, 14, 18, 19,* and *29*.) Athanasius figured prominently in the history, playing of course the role of the villain. Passages in the history served as first drafts for a separate treatise, *Paradoxical Questions Concerning the Morals and Actions of Athanasius*, which dated from near 1680, and which McLachlan has published in his volume based on the theological papers in King's College. (*Keynes MS. 10. Cf.* a MS. in the Clarke Library, Los Angeles.) In the history of the church, Newton also devoted considerable attention to the rise of monasticism, which he saw as a further dimension of Athanasius' baleful influence on the church. He interested himself in stories he found in the accounts of the early monks of their temptations and unclean thoughts. He collected every such story he could find, and rather solemnly he lectured the monks, who were alas beyond the possibility of profiting from his advice, on how to avoid such thoughts:

> Lust by being forcibly restrained & by struggling with it is always inflamed. The way to be chast is not to contend & struggle with unchaste thoughts but to decline them, keep the mind imployed about other things; for he that's always thinking of chastity will be always thinking of women & every contest with unchast thoughts will leave such impressions upon the mind as shall make those thoughts apt to return more frequently. (Clarke Library MS. of *Paradoxical Questions*.)

A new dimension to Newton's interpretation of the prophecies appeared about 1680, the role of Jewish ceremony as a figure or type. (*Yahuda MS. 2.4*).) Newton may well have owed this insight to Henry More, with whom he discussed the prophecies at this time. Wherever it came from, he approached it in his usual manner with a thorough study of Judaism. He read what he called the Talmudic scholars as well as the writings of Josephus, Philo, and Maimonides. (*Yahuda MSS. 13.2* and *28.5*.) The Jewish theme, which entered his interpretation at this time, remained a constant feature in it and appears prominently in the published *Observations*.

Jewish worship led him into the question of the exact dimensions of the Jewish temple, the best description of which he found in *Ezekiel* xl–xliii. As frequently happened with Newton, the problem of the temple took on a life of its own. In the end, he reconstructed the text of *Ezekiel* from a collation of all the versions available to him, including Hebraic ones, and from the necessities inherent in his plan of the temple which he drew from the texts and used as a device to criticize individual versions. (*Babson MS. 434*; *Yahuda MSS. 2.4*, ff. 17–49; *14*, ff. 1–8, 32–3; and *28.4*.) The plan of the temple conducted him further to the question of the length of the sacred cubit of the Jews, which he set at this time between the limits of 25 1/5 and 26 1/4 Roman inches. (*Yahuda MS. 2.4*, f. 40.) Later he confined the limits more narrowly.

Jewish ceremonies were not the only new theme to appear in Newton's interpretation of the prophecies in the 80's. The point of his interpretation also shifted from its earlier explicit identification of trinitarianism as the Great Apostacy to the concealment of its anti-trinitarian animus (though not its elimination) by an overt attack on Catholicism. (*Keynes MS. 5*, ff. I–VI, 1–6; *Yahuda MSS. 2.5a*, ff. 1–4; *3*; *9*; and *10.2*.) As far as I can tell, his theological opinions did not undergo any corresponding change. Quite the contrary, Newton's Arianism appears to have remained unchanged to the end of his life. An obvious interpretation of the change presents itself. With the 80's, the likelihood of the succession of James II became ever stronger. Faced directly with the Catholic enemy, Newton may have decided that he should find common ground with the rest of English Protestantism from which he had, in my opinion, consciously alienated himself. Two other changes in his interpretation which must be taken into account may possibly complicate this suggestion. On the one hand, Newton began to extend the time before the second coming. By the yardstick of the internal chronology of 1260 years and the time at which they began, Newton moved the second coming from the 19th into the 20th century; eventually he weould displace it still farther into the future. At the same time, the passion so prominent in the early treatise simply disappeared, as his exposition changed rather quickly toward the colorless narratives that make up his published *Observations*. Possibly we can relate these changes to the political realities of England in the early 80's. On this reading, the elimination of passion could have been the result of strict self-discipline as Newton strove to purge his interpretation of every element apt to compromise the struggle against the common enemy. The extension of the time before the last judgment could have represented a growing cynicism about the possibility of changing the existing order of society and a reluctant willingness to live with it. Newton's motives are wholly speculative since he did not spell them out. It is not speculation that his interpretation of the prophecies underwent the changes I have described. Neither is it speculation that Newton the archheretic led the resistance in Cambridge that preserved the Anglican university from James' endeavor to catholicize it, and that not long thereafter the man ready in the 70's to surrender his fellowship before he receive the mark of the beast accepted a rich appointment in London as his reward.

In his writings on the prophecies in the 80's, Newton also began for the first time to spell out the details of the second coming as he understood them. (*Yahuda MS. 9.2*, ff. 123–70.) He did not envision a cataclysmic destruction of the world natural but rather the end of the existing world political, that is the destruction of the unchristian regimes of the apostacy and the beginning of a true Christian society. In his description of the second coming, we can see further the implications of Newton's Arianism for his interest in the prophecies. One historian of Christian thought speaks of a tension, during the early centuries of the faith, between already and not yet. The establishment of the church resolved the tension on the side of already. With the coming of Christ, divine grace entered human affairs, and the permanent means of appropriating that grace in individual life exist in the established church. For Newton the Arian, the original tension was resolved on the side of not yet. The first coming of Christ was a relatively minor event. The great day of history, when the risen Christ, exalted for his obedience to sit at the right hand of God Almighty, shall come in his glory, remained still in the future.

The implicit denial of the decisive importance of the first coming of Christ was also the theme of a new theological work that Newton understook in the mid 80's, *Theologiae gentilis origines philosophicae*. (*Yahuda MSS. 13.3, 16, 17, 33*, and *41*; *Keynes MS. 146*; the MS. in the American Philosophical Society Library.) Though it was the most important theological treatise that Newton ever wrote, the very existence of the *Origines* has remained unknown until the present day. Minor aspects of it, which crept into the opening paragraph of the first draft of the *Principia's* concluding book (what is called the *System of the World*) and into the concluding paragraph of Query 31, have puzzled students of Newton who have not had their source available. Like his interpretation of the prophecies, the *Origines* took themes familiar to the age and converted them to an argument subversive of orthodoxy. According to Newton, all the ancient peoples worshipped the same twelve gods, who were identified with the seven planets, the four elements, and the quintessence. As gods, the twelve were deified ancestors, in fact Noah, his children, and his grandchildren, from whom all peoples descended. The worship of ancestors had not been the original religion of mankind, however. Noah and his sons had worshipped the one true God, the creator of the universe. They had worshipped him in temples, like the vestal temples in Rome, with a fire in the center, representing the heliocentric universe. But men possess an innate tendency to superstition and idolatry. Everywhere the original pure worship of God had been corrupted. The initial corruption of religion had happened in Egypt, and the rest of the ancient peoples had learned it from that source. As the Egyptians had Egyptianized Noah and his children and given them the names of their own heroes, so other peoples had also striven to magnify themselves by deifying their own ancestors. Hence the bewildering variety of ancient Gods, though Newton insisted that a unifying pattern stood behind them. Through the ages, God sent repeated prophets to restore the original pure worship. Despairing of mankind as a whole, he chose the Jewish people and sent Moses and the other Jewish prophets to them, all preaching the return to worship of the one true God and to love of neighbor, the two basic tenets of true religion. God eventually despaired of the Jews, and sent Christ as a prophet to the gentiles. Christ added nothing to the true religion that had been known to the sons of Noah. Like the Jewish prophets before him, he came not to institute a new religion but to call men back to the original one.

The interpretation of the *Origines* presents some complexities. On the one hand, Newton accepted the Biblical account of the creation and the early history of the human race. The Biblical account in this case was the Jewish account, however, not a Christian one, and Newton further confirmed it with non-Biblical sources such as the list of the post-Adamic kings by Berosius the Chaldean and the chronicles of the Syrian, Sanchoniathon, both of which he treated on a plane of equality with the Biblical story. What was more important was his deflation of the role of Christ in human history. Christ added nothing new to true religion. He came only as one prophet in a succession of prophets to call men back to the original pure worship. Newton further suggested that the true religion was known to all men via the study of nature. He found evidence that the same form of temple with a fire in the center, like the Jewish temple and the Roman vestal temples, had been used by all early peoples. This temple represented the universe, the true temple of God. By the study of nature, as Newton later put it in the 31st Query, men can recognize the creator and understand their duties toward him. Such is all that Christ re-affirmed. Thus the *Origines* seems to me to repeat the central message of Newton's interpretation of the prophecies – to wit, that the Bible was not a revelation of mysteries beyond human reason, mysteries unto life eternal, but an historical account of God's dominion, meant to demonstrate his omnipotent power to men, as nature demonstrates his infinite wisdom. Without mentioning the prophecies, the *Origines* gave further expression to Newton's conviction that they were the central element in the Bible.

It is worth noting the role in human history that the *Origines* assigned to Egypt, since it has some relevance to the matter presently much discussed among Newtonian scholars of the influence of the Hermetic tradition on Newton. According to the *Origines*, Egypt was the source of the "astronomical theology" of the ancient world. But this theology was error, not truth. More than error, it was the very embodiment of the fundamental human sin, the tendency to idolatry. Moreover, the corruption of true natural philosophy had followed in the wake of the corruption of true religion. Whereas the vestal temples of original worship had represented the sun centered universe, geocentric astronomy and the natural philosophy that accompanied it had followed the new theology. Ptolemy, be it noted, was another Egyptian. What is more important theologically, the idolatry of gentile theology, with its worship of men as gods, was identical to the idolatry of the Christian "homousians" (as Newton called them), a further legacy of Egypt via the agency of Athanasius and the early monks.

One significant conclusion which emerges from the study of Newton's theological manuscripts is the realization that virtually all of his creative life belonged to the years before his move to London in 1696. It was not merely Newton's achievements in mathematics and science, including alchemy, which stemmed from this period, but also the central ideas of his extensive theological effort. As far as I can tell from the manuscripts, the *Principia* interrupted his theological study, which together with alchemy had dominated his life for approximately fifteen years from the early 70's to the mid 80's. Although he was ready to discuss his conclusions with trusted intimates, to write out his view of the trinitarian corruption of Scripture for John Locke, and to begin a remodeling of the *Origines* which converted it into a chronology of the ancient world, I cannot find any significant number of theological manuscripts from a period of about two decades following Halley's visit in 1684. Only toward the end of the first decade of the 18th century – a date once again based primarily on my judgment of his hand – did Newton return seriously to theology. Furthermore, I am not able to find anything new in the papers from his old age. They concerned themselves entirely with working over ideas that belonged originally to the 70's and 80's. From the later period came most of the theological papers in the Keynes collection at King's, so imperfectly published by McLachlan. Pieces such as the "short schem of the true Religion" and "Irenicum" (which especially is mutilated in the McLachlan volume) appear to me to represent a melding together of Newton's continuing Arianism with the ideas of the *Origines*. (*Keynes MSS. 3, 6, 7, 8, 9*, and *11*.) He also took up the *Origines* once more, purged it of flagrant heterodoxy, and turned it into the pallid *Chronology of Ancient Kingdoms Amended*. He returned again to the prophecies in his old age, concentrating now more on *Daniel*. (*Yahuda MSS. 6, 7* and *8*; *Keynes MS. 5*, ff. 7–138.) He moved the date of the second coming back even further, to the 23rd or 24th century, and further obscured the original point of his interpretation to the extent that it can be seen only dimly in the published work even when one is familiar with the early treatise. Whatever his expectations of deliverance had been in the 70's, a certain cynicism of old age appeared to dominate his outlook – perhaps a recognition that the revolutionary regime which he had helped to install could not by any stretch of the imagination be mistaken for the kingdom of God on earth. Whiston recorded that Newton was wholly pessimistic about the immediate future.

I have tried to indicate the correlation of Newton's theological activity with his scientific career. Because of his immense importance in the history of science, any product of his pen that helps to illuminate his thought is significant. Had he not been a great scientist, however, it seems wholly unlikely to me that anyone would bother to study his theological manuscripts. If that is in fact correct, it is proper to conclude with a few general remarks about the relations of Newton's science and his theology.

Having studied the entire corpus of his theological papers, I remain unconvinced that it is valid to speak of a theological influence on Newton's science. I say specifically "theological influence," not "religious influence." The second can, I believe, be readily shown and is generally admitted. A theological influence, by which I mean the influence of Newton's central Arian position and his allied view of the prophecies, is another matter. As I indicated earlier, perhaps we can find the source of the God of the General Scholium in his Arianism. It is not clear which came first, however, his view of God or his Arianism; and even if we grant the influence, we remain still on a plane of high generality from which it is difficult if not impossible to demonstrate an influence on some concrete element of his science.

I myself am inclined to examine the relation from the other side. At the end of the 17th century, theology was the study with a firmly established, long dominant role of European civilization, a role then beginning to be challenged by the early success of modern science. It appears to me that we are more likely to find the flow of influence moving from science, the rising enterprise, toward theology, the old and (as we know from hindsight) fading one. Several recent articles and dissertations have insisted on the traditional elements in Newton's religion. With reference partly to my own earlier study of science and religion, they have urged that we not make Newton into too much of an 18th century figure. Central to the argument of these articles is the role of the Bible in Newton's religious thought. I agree without reservation that the Bible was to Newton the word of God in the fullest sense of that phrase. Nevertheless, I find it impossible to reduce his religion – or for that matter his concept of the word of God – to a traditional pattern. One must, I believe, start with and continually insist upon the central fact of Newton's Arianism. When we recall the role of Arianism in the ancient world and the role of its offspring, unitarianism, in the modern world, it becomes impossible, in my opinion, to ignore the influence of science upon Newton's religious views. I do not refer to what I myself would call Newtonian science. By Newtonian science, I mean primarily the science of the *Principia*, exact quantitative science based on the concept of attraction. Newton's theological position was formulated in the 70's before he had conceived the *Principia*. In speaking of the influence of science on Newton's religion, I refer then to more general conceptions basic to the whole scientific revolution, especially a new criterion of truth and a new locus of intellectual authority which ignored the claims of revelation. In Newton's case, the impact of these conceptions is most evident in his *Origins of Gentile Theology*, which I consider his most important theological work. The *Origines* asserted that the correct concept of the universe and the undefiled form of true religion are both found in the study of nature. What then was the Biblical revelation? It was to Newton God's revelation of his plan for human history; hence the central importance of the prophecies. As God demonstrates his omniscience to the student of nature, so also he demonstrates his dominion through the prophecies. Moreover, Newton's interpretation of the prophecies immeasurably reduced the role of Christ, which was central to the traditional Christian view of history. The first coming of Christ, far from being the central event of human history, merely repeated a pattern already established. The religion he established, a restoration of the original pure worship known to the sons of Noah, had through the worship of Christ as God been corrupted into idolatry, as that original worship had been repeatedly in earlier ages. Is this making Newton too much an 18th century figure? I can only say that on my calendar, the 18th century immediately follows the late 17th, and it does not appear implausible to me to find its roots there. And for that matter, Newton was an 18th century man, who was still alive when Voltaire visited England, and was writing theology at the time.

Indiana University

A CHECKLIST OF NEWTON'S THEOLOGICAL MANUSCRIPTS

Yaduda MS.	1 :	the early treatise on *Revelation* (70's)
Yahuda MS.	2.1:	part of the early treatise on *Revelation* (70's)
	2.2:	history of the church in Latin (c. 1680)
	2.3:	history of the church in Latin (c. 1680)
	2.4:	the role of Jewish ceremony as a figure or type in prophecy (early 80's)
	2.5:	three separate portions, in Latin, mostly on the history of the church; 2.5c, the third, on versions of *Revelation* (c. 1680)
Yahuda MS.	3 :	on *Revelation*, in Latin (early 80's)
Yahuda MS.	4 :	variant readings of *Revelation* (c. 1680)
Yahuda MS.	5.1:	part of a life of Christ (early 18th century)
	5.2:	notes on Irenaeus and Justin (late 70's)
	5.3:	notes on Athanasius (late 70's)
Yahuda MS.	6 :	on the interpretation of the prophecies (18th century)
Yahuda MS.	7 :	chaotic mass of short, disconnected pieces on the prophecies (18th century)
Yahuda MS.	8 :	fragments on the prophecies (18th century)
Yahuda MS.	9 :	on *Revelation* (1680's)
Yahuda MS.	10 :	fragments on the prophecies (70's and 80's)
Yahuda MS.	11 :	the beginning of a history of the church, in Latin (late 70's)
Yahuda MS.	12 :	part of a history of the church, in Latin (late 70's)
Yahuda MS.	13.1:	notes from Irenaeus (early 70's)
	13.2:	notes on Judaism (c. 1680)
	13.3:	notes related to the *Origines* (80's)
Yahuda MS.	14 :	a large number of disconnected fragments with notes on the fathers, on the prophecies, and on the church (mostly 70's)
Yahuda MS.	15 :	a mass of disconnected fragments, much of it on the original creed and faith (mostly post 1710)

Yahuda MS. 16 : *Theologiae gentilis origines philosophicae* (80's)
Yahuda MS. 17 : *Theologiae gentilis origines philosophicae* (80's)
Yahuda MS. 18 : short fragment on the history of the church (80's)
Yahuda MS. 19 : a treatise on the history of the early church (late 70's)
Yahuda MS. 20 : Latin version of first part of *Two Notable Corruptions*, not in Newton's hand
Yahuda MS. 21 : sermon on a text from *2 Kings* (80's)
Yahuda MS. 22 : professions of faith by early Councils, copied by Humphrey Newton
Yahuda MS. 23 : a short treatise interpreting *Revelation* in Wickins' hand, not Newton's composition
Yahuda MS 25 : Chronology (mostly post 1710)
Yahuda MS. 26 : Chronology (mostly post 1710)
Yahuda MS. 28 : fragments on the prophecies and Judaism (late 70's and early 80's)
Yahuda MS. 29 : a fragment on church history (late 70's)
Yahuda MS. 33 : a single page of notes related to *Origines* (80's)
Yahuda MS. 39 : a single sheet on *Revelation* (70's)
Yahuda MS. 41 : a chapter from the *Origines* (90's)
Keynes MS. 1 : part of a Latin treatise on *Revelation* (late 70's)
Keynes MS. 2 : Newton's theological notebook (mostly 70's)
Keynes MS. 3 : *Irenicum* and related papers (mostly post 1710)
Keynes MS. 4 : notes from Petau on doctrine in the 4th century (mid 70's)
Keynes MS. 5 : ff. I–VI, 1–6, on the interpretation of the prophecies (c. 1680) ff. 7–138, a treatise on the prophecies (18th century)
Keynes MS. 6 : seven points on religion (post 1710)
Keynes MS. 7 : 'A short schem of the true Religion' (post 1710)
Keynes MS. 8 : twelve articles on religion (post 1710)
Keynes MS. 9 : sheets on true religion (post 1710)
Keynes MS. 10 : 'Paradoxical Questions concerning the morals & actions of Athanasius & his followers' (80's)
Keynes MS. 11 : twenty-three queries on the word "homoousios"
Keynes MS. 146 : a chapter related to the *Origines* (90's)
New College MS. 361.1 (Bodleian): Chronology (mostly post 1710)
New College MS. 361.2 (Bodleian): Chronology (mostly post 1710)
New College MS. 361.3 (Bodleian): Chronology (mostly post 1710)
New College MS. 361.4 (Bodleian): mostly copies of "*Two Notable Corruptions*" and related papers. (90's)
MS. Locke c. 27, f. 88a (Bodleian): a tabular form of Newton's interpretation of *Revelation*
Babson MS. 434 : the temple of Solomon (80's)
Babson MS. 435 : a sheet on the chief rulers of the synagogue on the back of a letter (1709)
Babson MS. 436 : on the theology of Athanasius, in Latin (70's)
Babson MS. 437 : portion of a sermon on a text from *2 Kings* (80's)
Babson MS. 438 : a discourse on true religion (post 1710)
Babson MS. 704 : a short fragment with a note on the history of the early church (70's)
Clark Library (Los Angeles) MS: paradoxical questions concerning Athanasius (80's)

Clark Library MS : notes from Cudworth (80's)
Humanities Research Center (University of Texas): a sermon on a text from *2 Kings* (80's)
Trinity College, MS. R. 16.38, ff. 436 and 438: notes from Cudworth and an ancient chronicle (80's)
Bodmer MS : contents and date unknown

J. E. McGUIRE

SPACE, INFINITY AND INDIVISIBILITY: NEWTON ON THE CREATION OF MATTER

I. INTRODUCTION

About 1668 Newton composed a metaphysical treatise now entitled *De Gravitatione et Aequipondio Fluidorum.*[1] Despite its stated subject, the work develops an account of the nature of space, time, matter and motion. Newton's immedicate motivation is a desire to refute what are for him the chief assumptions of Descartes' natural philosophy. He consequently embarks on an examination of the central doctrines of the latter's *Principia Philosophiae*. The result is the systematic establishment in *De Gravitatione* of the basic principles of Newton's own metaphysics of nature.

I wish to concentrate on Newton's view of how space and matter relate. While this view differs in certain details from the theory later presented in his *Principia*, in its account of the individual nature of space, it sets the framework for Newton's mature thought on the subject.[2] It is therefore important to understand the metaphysics of this early treatise. But it is also important to appreciate the extent to which Newton's conception of space and matter emerges from a rejection of Cartesian principles,[3] while at the same time reflecting some important doctrines of earlier natural philosophy, especially the Epicurean notions of indivisible minima and the continuity of magnitude.

I shall begin by outlining Newton's general intellectual orientation. After rebutting Descartes' conception of motion to his own satisfaction, Newton turns to the main subject of his discourse – the nature of space and its relation to matter. He is concerned in particular to reject two views concerning the nature of corporeal substance. The "vulgar notion" which conceives it as an unknowable substratum in which properties and accidents inhere; and the Cartesian conception that its essential nature is to be extended in length, width and breadth.[4] Newton rejects the first conception because he considers it to be unintelligible. His reasons are these. If the reality of substance is known only through its properties, what is this 'thing' that they are said to inhere in? On this conception, a substratum is an undifferentiated "thing" at once distinct from its instantiated properties. And since it lacks features which in themselves are perceivable, there can be no separate characterization

Z. Bechler (ed.), Contemporary Newtonian Research, 145–190.

of that in which perceivable properties inhere.[5] Newton holds, reasonably enough, that only substances have substrata. He seems to assume, however, that those who advocate substrata also believe them to provide the only account of property instantiation. In any event, in Newton's view, generic extension is in itself a thing which is distinctly conceivable. And in a statement which is effectively the subject of analysis for the present paper, he says that "substantial reality is rather to be ascribed to attributes [e.g., extension] of the kind which are real and intelligible in and of themselves, and which do not need to be inherent in a subject".[6] This ontology is to be preferred to an unknowable substratum of whose nature we lack a positive idea.

Newton's worry about Descartes' conception of corporeal reality is twofold. On the one hand, he correctly takes Descartes to hold that the essential nature of corporeal substance is to be extended tri-dimensionally, so that in reality the natures of matter and space are indistinct.[7] This conclusion is unacceptable to Newton. On the other hand, he rightly construes Descartes as holding that being extended, *per se*, is necessary and sufficient for anything to be material in kind. But by imposing on Descartes' position his own view of extension as that which is infinite, uncreated and eternal, Newton concludes that Descartes' conception of matter leads to atheism.[8] For certainly if Descartes' conception of extension is interpreted along Newtonian lines, it indeed leads to the conclusion that matter exists as an knowable reality independently of knowing how it relates to God's nature.[9] And Newton wishes to reject this conception of matter in order to advance a theory of corporeal nature which is intelligible, not just because it avoids the substratum theory and Cartesianism, but because it shows how matter depends ontologically on Divine nature. In other words, Newton wants an account of the nature and origin of matter in virtue of which it cannot be understood without also comprehending the implications of Divine omnipotence. The extent to which Newton's concern is in fact motivated by his confusing Cartesian extension with his own conception of uncreated extension is considered below.

Newton's theory of matter and its relation to space is intended to meet these alleged difficulties in the Cartesian and substratum accounts of material reality. And Newton develops it by considering the manner in which God could have created matter *ex nihilo*. He begins with the doctrine that infinite extension – a defining feature of space – is "eternal, infinite, uncreated, [*increata*] and uniform throughout, not in the least mobile, nor capable of inducing change of motion in bodies, or change of thought in the mind".[10] Furthermore, he holds that infinitely extended space is co-eternal with God's uncreated existence. For "if space had ever not existed, God then would have been nowhere, and hence he either created space later, where he himself was not, or else, which is no less absurd to reason, he created his own ubiquity".[11] Thus, according to Newton's conception of the nature of Divine presence, the co-eternal existence of God and space is an integral part of the scheme of things. Moreover, Newton conceives both God and space to be infinite in virtue of their respective natures. This does not mean, however, that infinite space embodies in its nature the intrinsic ground of its individual existence. It exists because God exists; for space is an immediate consequence of the eternal presence of Divine being.[12] God neither creates nor conserves its existence, anymore than he creates and conserves his own ubiquity.[13]

It is in terms of this conceptual framework that Newton advances his hypothesis regarding the creation of matter. Divine power conditions omnipresent figures in the uncreated extension of space, so that each is a causal ground necessary and sufficient to produce those preceptions of properties which the mind attributes to the nature of corporeal phenomena.[14] Before discussing this hypothesis in detail, it is necessary to consider Newton's ontology of space in its own right and in its relationship with Epicureanism. In the last section of the paper I shall analyze Newton's critique of Descartes' ontology.

II. THE NATURE AND STRUCTURE OF INFINITELY EXTENDED SPACE

Newton decided early that extended space is real and in fact infinite. He unequivocally states this belief in his first scientific notebook the *Questiones quædam philosophicæ* (1661–65):

> If extension is indefinite onely in greatness & not infinite, y^{n} a point is but indefinitely little & yet we cannot comprehend anything less. To say y^{t} extension is but indefinite (I meane all ye extension wch exists & not soe much onely as we can fansy) because we cannot perceive its limits, is as much to say God is but indefinitely perfect because wee cannot apprehend his whole perfection.[15]

Thus extension – a defining characteristic of space – is infinite. Moreover, Newton denies Descartes' claim that we only understand extension, *per se*, to be indefinitely great. In his view, we can understand *that* extension is infinite, even though the mind is unable to form a completely positive idea of the infinite actuality of its perfection. Only God can do that. This conception of space is argued for in detail in *De Gravitatione*. And so, too, is the claim that space possesses an inherent ontological structure. The earliest evidence for Newton's belief in the ontological structure of spatial extension is also found in the *Questiones*. However, the analysis required for a full account of Newton's views in these early writings goes beyond the present paper. I shall, therefore, consider his views as they are developed in *De Gravitatione*, referring to the *Questiones* where appropriate.

Newton begins his account of the ontology of space by defining extension as "the uniform and unlimited stretching out of space in length, breadth and depth"; for "space everywhere extends to infinity. For we cannot imagine any limit anywhere without at the same time understanding [*intelligamus*] that there is space beyond it". Thus, the intellect understand that space is in fact infinite in the sense that neither void nor anything other than space can possibly exist beyond it and hence limit it. This contrasts with the possibility that something other than a finite being can in fact exist beyond the limits of its extremities. And since Newton holds that that which is unlimited is fully actual, he at once links actuality of infinity with perfection of being.[16] Thus, instead of the Aristotelian notion that the infinite is an imperfection where always something remains, Newton holds that the infinity of space is a perfection beyond which there is nothing. When these ideas are grasped correctly, a number of space's important features, all of which are embodied in its defining nature, can be considered. In Newton's view, these features show both that space *is* something and that it possesses a definite nature. His account of its general structure is of considerable interest. Because of its interconnected claims, it will be helpful to quote the passage at length:

Everywhere space [*spatium omnifariam*] can be distinguished into parts whose common boundaries [*terminos communes*] we customarily call surfaces; and these surfaces can everywhere [*omnifariam*] be distinguished into parts, whose common boundaries we call lines; and again, these lines can everywhere be distinguished into parts which we say are points. And hence surfaces do not have depth, nor lines breadth, nor points any dimension, unless you say that bordering space [*spatia contermina*] mutually penetrate as far as the depth of the surface between them, namely, what I have said to be the boundary of both or the common extremity [*extremitatem communem*], and so for lines and points. Furthermore, spaces are everywhere contiguous to spaces, and extension is everywhere placed next to extension, and so there are everywhere common boundaries to continuous parts; that is, there are everywhere surfaces as dividing boundaries to solids and everywhere lines in which the parts of the surfaces touch each other; and everywhere points in which the continuous parts of lines are joined together. And hence there are everywhere all kinds of figures, everywhere spheres, cubes, triangles, straight lines, and those of all shapes and magnitudes, even though they are not disclosed to sight [*non ad visum delineatae*]. For the material delineation of any figure is not a new production of that figure with respect to space, but only a corporeal representation of it, so that what was first [*prius*] insensible in space [*in spatio*] now appears to the senses to exist. For we firmly believe that space was spherical before the sphere occupied it, so that it could contain the sphere; and hence as there are everywhere spaces that can adequately contain any material sphere, it is clear that space is everywhere spherical. And so of other figures.[17]

There is much here which requires comment, and it will, in large measure, occupy the remainder of the analysis. Newton's central claim is that extended space *actually* contains three-dimensional solids of all shapes and magnitudes, whose properties can be characterized in geometrical terms. Put otherwise, volumetric figures are said to be embedded in space, whether or not any material thing really occupies any of its given regions. In Newton's view, space therefore possesses constituents which determine its actually inherent structure. Given this ontology, Newton holds that there are dimensions embedded in space's structure which, when characterized geometrically, pertain to the distances between its variously embedded regions. Consequently, in Newton's view, 'distance' is not merely a two-place predicate which characterizes a certain disposition among material objects. Rather, the distance (at a given time) between two objects, is the actual distance between those parts or regions in space which the objects in fact *occupy* at that time. This conception of distance is embodied in the opening definitions of *De Gravitatione*. Definition I states that "Place is a part of space that something fills evenly;" while the second definition states that "Body is that which fills place".[18] Thus the actual distance between objects, is the distance between the different parts in space that they in fact occupy. And since places are embedded in space, there are in fact distances between them (unless they are adjacent), which do not depend on their being occupied.

At this point an important question arises. Does Newton think that space possesses a natural metric? The query is proper enough, if like Newton one holds that distance is a real element in the ontology of space. Although an answer will be constructed from what Newton says in *De Gravitatione*, some background from the *Questiones* is necessary. There are a number of entries from the early manuscript in which Newton argues that mathematical points, which are dimensionless and indivisible, are not the sorts of entities which can ground the nature of a real extension, whether that be the dimensions of corporeal objects or an actual magnitude in space.[19] He naturally enough concludes that the dimensions of any object, which is truly extended, cannot be generated by an infinite number of mathematical points. Rejecting an ontology of points, Newton argues, in Epicurean fashion, that any extended object is composed of minima – partless units which in fact possess indivisible magnitude. With corporeal objects, these are atoms; with space, it is the infinitude of least parts that compose its ultimate structure, since "an infinite number of extended parts (& y^t least parts of quantity must be extended) make a thing infinitely extended."[20] So in Newton's view, the extended structure of space is constituted by infinitely many partless minima which are its least and indivisible units, but not in fact composed of mathematical points.

In Newton's world, furthermore, there is no room for mere potentiality in any form. All existing things, space included, require a structure of ultimate, indivisible and actual parts. And unlike corporeal things, the nature of space is such that it cannot be split into its constituents. The problem remains, however, as to how these minima are reconcilable with the claim that space is *actually* infinite; for though infinite in number, each minima is finite in itself.[21]

In terms of this ontology, Newton argues in the *Questiones* that the length of any actual distance is composed of a number of real and determinate minima which are its ultimate units of measurement, and which inhere in the structure of space itself. That is, the ontological constituents of any extended portion of space are the least parts of the same indivisible size that together cover its distance-interval *partes extra partes*. So as early as the *Questiones*, Newton is concerned with a significant problem; namely, how to provide an explanation of metric properties in terms of non-arbitrary minima intrinsic to the nature of space itself. What he is moving towards, but not to his complete satisfaction at this stage, is a mean of conceiving the length of a given distance-interval in terms of the ratio of its intrinsic minima to those of another.[22] In other words, in Newton's view, the metrical properties of space are embedded in its structure; measuring devices are merely an epistemic means of discovering them.

These sorts of questions continue to occupy Newton in *De Gravitatione*. But straight-away, a difficulty confronts us in the lengthy passage above. Why does Newton state that points have no dimension, lines no breadth and surfaces no depth, only to say immediately "unless you say that bordering spaces mutually penetrate as far as the surface between them, namely, what I have said to be the boundary of both or the common extremity, and so for lines and points". More particularly, why does he suggest that the bounding surfaces of the figures inherent in space have minimal depth?

I think that the statement following the phrase "unless you say . . . ", cannot be construed as merely offering a possible view which someone might want to embrace. On the contrary, it contains what Newton takes to be the case concerning surfaces which are common boundaries of the figures which everywhere inform space. In other words, far from merely offering a possible interpretation of the nature of points, lines and surfaces, Newton is claiming, albeit rather awkwardly, that he understands these entities in a sense which differs from their conception in geometry itself.

It seems that there are two reasons why this is the case. Newton claims that points have dimension, lines breadth and surfaces depth, because his figures are entities inherent in an omnipresent space, and consequently they have intrinsic extension and real extremities. And since he is dealing with the actuality of magnitude, indivisible points, breadthless lines, and surfaces without depth are not proper elements of the ontology. Second, Newton is concerned, not surprisingly, with the problem of continuity as it relates to the successive juxtaposition of these figures, existing as they do in the reality of space itself. So while indivisible points, and lines and surfaces which lack real extension can be conceived in geometry, the extremities (surfaces, edges and points) of the volumetric figures cannot be so conceived. After all, Newton denies that 'being indivisible' and 'being without magnitude' entail each other, and would agree with Epicurus that Aristotle has not shown that any indivisible unit whatsoever must lack size:[23] the existence of indivisible magnitudes can be demonstrated (*Questiones*, f 89^r, f 119^r). Consequently, Newton's ontology of space requires an account of the continuous which differs from the geometrical conception of magnitude as indefinitely divisible, and also from the subtle Aristotelian ontology in which the continuous is that which is everywhere and potentially divisible to infinity, though not, of course, actually divided everywhere at the same time.[24]

But Newton's conception of space embodies two distinct ontological levels at which the problem of spatial continuity arises; (a) the manner in which space's volumetric figures are said to be continuous, and (b) the nature of the continuity of the indivisible minima which (as I shall argue) form the ultimate structure of space in *De Gravitatione*. Consider the volumetric figures first. In virtue of their intrinsic existence in space itself, these figures are indivisible, three-dimensional shells whose bounding surfaces possess minimal depth. Moreover, they are of such sizes and shapes that, when God endows them with the capacity to produce representations of material reality in the perceiving mind, they provide the extremities necessary for demarcating visible and corporeal things from one another. Volumetric figures are thus of macro-size. And since their three-dimensional pattern of shapes is such that they fill the whole of infinite space without any interstices remaining (a clear geometrical possibility), the continuity of these spatial figures is raised in an interesting way.

To appreciate this, and also the continuous nature of indivisible spatial minima, a brief statement of Aristotle's conception of the continuous is called for. At *Physics Z* (231a21–231b6), he makes two observations concerning the continuous. Anything continuous cannot be made of indivisibles such as points; for things can only be said to be continuous just in case their "extremities are one". But with respect to indivisible points, it is impossible to distinguish extremities. Nor can a continuous line, for example, be compose of points in contact. This notion demands, says Aristotle, that contacts be whole on whole, part to part, or a part with a whole. Since indivisibles cannot be said to have parts, contact between them must be whole to whole. But then, they are coincident and indistinguishable one, and consequently, cannot form a continuum; that requires parts differing from one another by a process of divisibility which makes them successively and separably distinct. Ironically enough, Aristotle's arguments only explicitly deny that points and instants can be arranged successively; as we shall see, an opening is left for someone like Newton to argue for the contact of indivisible minima of extension. The result is not a true continuum in Aristotle's sense, but it is none the worse for that.

Now with respect to Newton's conception of the continuity of macro-figures in space, it is essential to understand his use of the notions of "common boundaries" and "common extremity". Bordering figures do not possess "common boundaries" in the sense in which two separate objects, when partially interpenetrating, can be said to share a boundary; but possess them in the sense that each individual pair of figures, conceived in succession, is intrinsically and unchangingly demarcated in space by *one* the *same* bounding surface. So when Newton says that bordering figures "mutually penetrate as far as the depth of the surface between them" that does not mean that they can actively interpenetrate in the manner of separate objects – how could they, given space's immutability and its inseparability from itself; rather, it means that these figures, conceived pair-wise, through the regions of space, are altogether united as one entity, in the sense that each pair entirely 'shares' a spatially inherent common boundary. There can be no doubt that Newton's conception satisfies the Aristotelian notion that things are continuous only "if their extremities are one", a notion which he no doubt has in mind. We therefore have a picture of infinite rows of volumetric figures present in all directions and at all angles to each other in the infinitude of space, figures which everywhere juxtapose and intrinsically over-lap in intricate three-dimensional patterns. But in all cases, they are continuously and immutably 'fused' at their common boundaries, and so arranged that it is impossible for interstices to exist anywhere in space.

While it is true that in *De Gravitatione* Newton never specifically mentions an ontology of indivisible minima in space, nor raises the question of how a natural metric is derived from them, the ommission should cause no surprise. After all, he is concentrating on space's ontological macro-structure as an essential part of his account of matter's creation. Furthermore, he has satisfied himself on the ontology of indivisible least parts in the *Questiones*, and knows that the views expressed there are compatible with the nature of space as he presents it in *De Gravitatione*. After all, only three or four years separate the composition of the two works. However, since the continuity of indivisible minima is important for understanding Newton's conception of the unity of space, related as it is to the ontology of Divine existence, it is well to give further consideration to how he probably conceived their successive juxtaposition at the time of writing *De Gravitatione*. The more so, since one of the treatise's main thrusts is to argue against Descartes's conception of relative motion for a conception of real motion with respect to intrinsic spatial distance. It would be surprising, therefore, if Newton were indifferent to the problem of conceiving the measure of motion in terms of minimal units inherent in space itself.

In the first place, it is clear that Newton must deny the Aristotelian claim that indivisible minima of any sort cannot be in contact. If Aristotle's theory of the continuum holds that anything continuous cannot be made of indivisibles, that claim must be met. The *Questiones* (as well as the general thrust of the reasoning in *De Gravitatione*) leave no doubt that Newton meets it by applying the Epicurean conception of the indivisible to his spatial minima. He thus conceives them as successively juxtaposed to each other, *partes extra partes*, and so arranged that they need not be in contact whole to whole or part to part. On the one hand, this means that Newton's indivisibles cannot be 'nothings' lacking extension, for these coincide by touching whole on whole; but on the other hand, they cannot be in contact part to part, since that means that they are in principle divisible.

These conditions are satisfied if it can be shown that the least parts are indivisible and partless extensions of minimum size which constitute absolute and non-arbitrary units of measures. Although the details of Newton's Epicurean style of argument against the divisibilists go beyond the scope of the present paper,[24] this is the sort of ontology that he attempts to establish in the *Questiones*. There, as we have already seen, he argues that spatial minima are not indivisible points without magnitude, but units of minimum extension. But he also says of his minima that they combine smallness with partlessness. He argues first that a minimum extension in space is so small that "a point cannot be put or conceived in this little space to divide it" (*Questiones*, f 119r). In answer to the objection that a "least extension is infinitely larger y^n a point & therefore can conteine it & be divided by it", Newton, in effect, argues that a minimum extension is partless, whether it is an atom or a spatial magnitude. So in a real yet small extension, a least distance is "from one extrem to another. It hath no inside, no midst, nor center but is it self all (center, inside & midst) to ye environing superficies". Thus you "cannot put a point wth in it because it hath no inside. Set a point uppon it & then it touches but its superficies"; points, therefore, "cannot make out a place for division" (*Questione*, f 119v. (Brackets and punctuation mine).

Although he does not make an explicit distinction between conceptual and actual divisibility, it seems that Newton is claiming that his minima are conceptually indivisible. They are in fact, he claims, without real parts. And given Newton's commitment to the real distinction (i.e., that the mind can conceive of differences in a quantity if and only if there is an intrinsic or appropriate ground in the individual itself, *Questiones*, f 88v), if the minima are in fact without differences, it is impossible that the mind can conceive of parts within them. In this respect, they clearly differ from the macro-figures. When materially delineated by God's power, these are perceptible magnitudes which can be conceptually differentiated by the mind. But each of the minima is everywhere the same and, being without intrinsic differentiation, Newton claims that each lacks a conceptual purchase for the would-be divider. His position, however, is not convincing. Although there is no procedure whereby space can be divided actually, it is not clear that Newton's arguments preclude the claim that half an indivisible minimum can be conceived of. In answer, Newton has either to deny that geometry applies to space (a radical move), or give more compelling arguments to shown that by their very nature his minima lack internal differences and can be distinguished from mathematical points. Nevertheless, since he allows in *De Gravitatione* that space possesses figures of "all shapes and magnitudes", there is nothing to prevent him assuming that it has minima of equal size and the same shape, a view which satisfies the principles of actualism and indivisibilism to which he subscribes.

Given that this is their nature, Newton spatial minima are continuous in the following sense: they are arranged successibley with "extension everywhere placed next to extension", each minima an indivisibly extended unit in absolute contact with the units adjacent to it. In fact, in the *Questiones* (f 91^v), Newton defines things contiguous as having their edges in the same place. As he puts it, "the surfaces of two bodies are one when they are contiguous, because in one place." With spatial minima, the requirement of absolute contact "in one place" is essential. For were it possible for intervening gaps to exist between any or all of the non-overlapping units, their combined length would be indeterminate, and thus the ontological basis for a natural metric would be destroyed. But are these indivisible units not in contact part to part, and hence liable to Aristotle's objections? Newton answers, in effect, that the extremities and edges of the spatial minima are not themselves parts. We can conceive their bounding extremities in terms of indivisible points, but these are not 'parts' which serve as a basis for division (being themselves nothings), let alone 'elements' which exist as proper parts of the indivisible minima themselves (*Questiones*, f. 119^v). So while the indivisible minima lack parts which can be divided into units of smaller dimensions, this does not mean that there is no distance from one extremity to the other, or that we cannot speak of one extremity as being opposite to another. Nor is there reason to deny that Newton's minima possess boundaries even though their extremities are not parts.

Newton's 'Epicurean' continuity raises questions which I need not deal with fully in the present context. There is a question about Aristotle's puzzles concerning indivisible motion. If an ontology of separate and indivisible magnitudes is supposed, it is necessary to posit indivisible distances and indivisible times. But on this hypothesis, an indivisible magnitude can only be said to have moved, never to be moving. Although he addresses this question directly in the *Questiones* (f. 92^v), Newton need not in *De Gravitatione*. As we shall see, the explicit theory of corporeal motion he advances, has only to consider the ontology of space's inherent macro-figures, figures which are its intrinsic and immutable constituents, but not the movement of separate atomic indivisibles which demand indivisibles spaces, distances, and times.

Another question pertains to Newton's conception of an inherent spatial metric. If it is assumed, naturally enough, that the indivisible units of measure are all of the same shape, as well as of the same minimum size, puzzles ensue. Suppose that each is square, and that the arrangement of shapes is such as to fill space without intervening gaps. If the rows of indivisibles are arranged at right angles to each other, the metric is determinate. This is not so, however, if they are arranged at varying angles, or, for example, if the indivisibles are spherical in shape. In the one case, there are problems of incommensurability, since every row cannot be expressed in whole numbers, whatever units are chosen; in the other, the shapes cannot be in absolute contact such that no interstices remain. Newton might argue that right angle coordinates reflect privileged directions in space itself, but this seems to be an unwarranted restriction. Unfortunately, he says little that gives any clue as to his view on the nature of complex metrical problems. And it may well be that these sorts of problems caused him to avoid explicit mention of intrinsic spatial parts in the *Principia*, even though he continues to hold that space is the autonomous ontological ground of inherent metrical properties.[25] But there is another important reason why Newton would probably conceive his minima as having the same size and shape: that arrangement satisfies the need for conceptual indivisibility. Anyone who posits atomic magnitudes must allow for innumerable shapes and sizes so as to account for manifest phenomena. However, if a large indivisible is superimposed on a smaller it is possible to conceive of parts in the portion that makes the difference between them. But if Newton conceives his spatial minima as squares of the *same* size, superimposition of this sort cannot occur, since they are immutable interlocked side by side in space.

At this point, it is necessary to give further consideration to the ontology of tri-dimensional macro-figures which constitute the rich variety of space's structure. An important notion in the *De Gravitatione* passage is that space contains "everywhere all kinds of figures – even though they are not disclosed to sight". The key term is 'everywhere' and it is one which Newton uses repeatedly. Moreover, it is a term which can easily be misunderstood on a casual reading. Newton states, for example, that "space is everywhere spherical" (*ubique esse spatium sphaerica*). The claim is then generalized to say that space is in fact everywhere informed by "all other kinds of figures." In claiming that space is 'everywhere' informed by different kinds of figures he does not mean that *somewhere* in the vast stretches of space each kind of figure is to be found. If this were his meaning, he could hardly maintain, as he in fact does in the passage, that figures of all sorts can be 'materially delineated' in each separate region of space. Rather, in Newton's view, space is said not only to contain figures of all kinds or sorts, but each of its ramified 'regions-within-regions' is said to contain all such figures. Accordingly, a complex picture of space's structure emerges. Any and every three-dimensional stretch in space has, as part of its inherent structure, a multitude of figures embodying all shapes and magnitudes. Thus all of space's encapsulated regions-within-regions etc., are so constituted that they contain ennumerable figures, each a magnitude which is compatible with a given region's intrinsic extension. Thus individual regions are simultaneously related in various ways;

they fully interpenetrate, they overlap, and their intrinsic figures are arranged successively. In sum, three dimensional space is constituded by an actuality of forms which are its omnipresent parts. And Newton has a ready answer to the ancient question: where do the objects of geometry exist? They are concreta that exist in infinitely extended space.

Newton's view of space as 'something' that possesses an inherently actual structure raises problems. Since extension is said to be primitive property of space, its inherent properties are primitively extended, too. In as much as all figures in a 'region' of space are extended, their individual dimensions are embedded in that region. However, each region contains a multitude of three-dimensional figures of varying dimensions. So despite their dimensions, they must in some cases overlap, and in others occupy the same 'place', i.e. coincide in the region itself. According to traditional thought, however, to say of something that it is dimensional, is to say that it is corporeal; for on this view dimensionality and corporeality are indissoluably connected.[26] On this account, extensional entities cannot be said to interpenetrate. They possess corporeal dimensions, and so can be said neither to overlap nor occupy the same place. But Newton is committed to holding that space's inherent figures do overlap and interpenetrate.

The traditional view is given added significance for Newton's position since it is reinforced by Descartes' theory that only corporeal substance is extended. For Descartes, the dimension of any given distance is a magnitude which is a proper mode of extension, since "any distance is a mode (modus) of extension, and thus cannot exist without an extended substance."[27] And the simple nature extension is itself the essential attribute of corporeal substance, in that being extended is necessary and sufficient for being a material thing. An important premise in Descartes' argument is that where there is extension, there must be some *thing* that is extended. Otherwise we are forced to say that *nothing* is extended, while at the same time admitting that non-being cannot instantiate properties.[28] This is an instant of a more general doctrine: if there is an attribute, it must be the attribute of some substance. Given these commitments, Descartes in effect argues as follows: extension is an attribute, and since there is only a distinction of reason between extension and space conceived as continuous magnitudes, where there is extension there is space. But an attribute cannot exist without a substance, and since extension is the essential attribute of corporeal nature, there is no extension without corporeal substance of which space is merely a determinate mode.[29] Thus neither an extended void nor a continuously empty vacuum can exist, since 'being extended' cannot pertain to what is incorporeal. So for Descartes, the real features of the external world do not warrant an atomistic distinction between the full and the empty.

What response can Newton make to these difficulties: to the traditional view that dimensions are possessed only by corporeal things; and to Descartes' denial that extension can be an attribute of incorporeal substance? But more to the point, can he establish his position against these objections in a way that does not obviously beg questions? In part, Newton's reply rests on claiming that there are senses in which extension, *per se*, is neither a substance nor an attribute; but he also wants to show that extended space can be distinguished from corporeal extension. As will become apparent, these claims are interconnected, and face certain difficulties.

Consider the first claim. Newton begins by saying that extension is neither a substance nor an accident "for it has a mode of being proper to itself which pertains to neither substances nor accidents."[30] Of the view that extension exists as an accident he says:

> . . . we can clearly conceive extension existing, as it were, without any subject, as when we imagine extra-mundane spaces or places empty of bodies; and we believe extension to exist wherever we imagine there are no bodies, nor can we believe that it would perish with body if God should annihilate the manner of being [of the latter] ; it follows that extension does not exist in the manner of an accident.[31]

So far as they go, Newton's remarks do little more than describe what he takes to be imaginable, and reduce to the claim that mind can abstract 'extension' from the various ways in which it pertains to material phenomena and so consider it in itself. Thus rather than *arguing* that extension has an actual mode of being which differs categorically from that of accident, Newton in fact presupposes what he must show. Accordingly there is nothing to convince Descartes (whom Newton has in mind here) that extension has a mode of existing which is intelligible independently of corporeal substance and its proper modes. Merely to deny, by means of thought experiments, that extension is a dependent modification of an underlying subject, is not to show that it possesses a particular form of positive being. Furthermore, Newton here treats items in the category of accident as being wholly *unrelated* to the natures of things in which they are present, and thus able to change independently of them. Unfortunately, it is in terms of this conception of accident that he erroneously construes Descartes' doctrine of extension as it pertains to corporeal reality. But in Descartes' view all corporeal modifications are essentially proper modes of extension, itself the essence of matter. Thus for Descartes it is contradictory to suppose that matter can exist independently of extension without the destruction of the latter.[32]

However, Newton's reasons for denying that extension falls under the category of substance (at least in some of the traditional senses) have more interesting implications. His general argument claims that extension is not a substance because it

does not subsist absolutely *per se*, but is, as it were, an emanative effect of God, [*sed tanquam Dei effectus emanativus*] and an affection of all being, and because it does not underlie [*substat*] proper affections of the kind that denominate substance, namely actions, such as thought in the mind, and motions in body.[33]

There are two reasons advanced. First, that extension is not substance because it is a consequence of Divine being and thus not an independent existent. And second, that it is not an affection that properly pertains to the nature of substantial reality. Here Newton has a basically Aristotelian conception of susbtance in mind: thus he denies that extension is a substance in the Aristotelian sense of an individual which is a source of change in itself or in other things, and which can be acted upon. In his view, extended space is deprived of causal properties, and capable only of the passive disposition of receptivity. This does not show, of course, that extension is not a susbtance in some other sense. The second reason turns on what Newton understands by the term 'affection', as well as on the sense in which he takes extended space to be "an affection of all being." A little background is necessary.

As I have argued elsewhere, Newton's thought in *De Gravitatione* is based upon the distinction between predicates of pure existence and predicates which pertain to the defining characteristics of actual existents.[34] Furthermore, as Newton uses this prediction schema, 'existence' itself is a pure predicate which refers to the sheer actuality of various sorts of existents, and is therefore not a *real* affection which contributes to the determination of their sort or kind. But extended space, as well as time, are also predicates of pure existence. They pertain to the *fact* that determinate sorts of thing exist at definite positions, or at successively different positions, in the nexus of space and time. Thus there are two sorts of affections: those which characterize natures *qua* natures, and those that pertain to the fact that individual natures exist *qua* existents together with their characterizing features.[35]

For present purposes I shall take this framework to be a decided element in Newton's thought, and proceed to his conception of 'affections' of being. Consider the following passage from *De Gravitatione*:

Space is an affection of being in so far as it is being . . . The same may be affirmed of duration: both are affections of being or attributes according to which the quantity of existence of each individual [being] is denominated as regards its amplitude of presence [*amplitudinem praesentiae*] and its perseverance in existence. So the quantity of existence of God is eternal in relation to duration, and infinite in relation to the space in which he is present; and the quantity of the existence of a created thing is as great, in relation to duration, as the duration since the beginning of its existence, and in relation to its amplitude of presence, as great as the space in which it is.[36]

The key notion is that embodied in the phrase "quantity of existence" especially as it is said to pertain to all existents. In the general case, Newton is working with a parallel sense in which time and space relate, respectively, to the duration and presence of existing things. Thus, as time is the measure of something's duration, so space is a measure of its "amplitude of presence." So far as it goes, however, the schema gives no grounds for claiming, as Newton wants, that space and time have a real mode of being in the sense of existing independently of other existents. Rather, space and time characterize anything's "quantity of existence" i.e. its duration and its present in space. Descartes in fact accepts the schema, and also consistently affirms that space and time are both *ens rationis*; that is, they are ways in which the mind comprehends something's continuance in existence, although the *esse* of space and time is to exist in the mind alone.[37]

But in Newton's view, extended space "is, as it were, an emanative effect of God;" or, as he alternatively puts it "an emanative effect of the primarily existing being, because when any being is posited, space is posited." Thus space exists because God exists, and is grounded in the fact that God *is* an eternal, uncreated and concrete being. So conceived, uncreated space (along with unending time) characterizes God's "quantity of existence," where to speak of something's "quantity of existence" is not to characterize a real attribute of its defining nature, but rather to speak of the sheer actuality of its determinate nature *qua* existent. And to be an existent is to be that which can be manifested in time or space, or in both. In the case of Divine presence, it means that God exists eternally with respect to unending time, and infinitely with respect to the uncreated extension of space; there is no time or place at which it is possible that God fail to exist. Thus far from being mere *entia rationis*, space and time are 'affections' of God's individual nature *qua* existent, where to say of something that it is an actual existent pertains to the fact that it exists with respect to the determinate nexus of space and time. So cosntrued, space and time are affections of pure existence. Thus, to say of something that it actually exists, is to say it is *there* in the objective order of space and time. As Newton puts it, the "Predicaments of *Quantitas Ubi & Quando* – are not qualities in the proper sense, but denote only places & times of the existence of beings contained in them."[38] When applied to

God's existence *Ubi* and *Quando* answer two general questions. Where does God exist?; 'everywhere', in respect to space. When does God exist?; 'always', in respect to time. This is just as to say of finite things that they are 'here' or 'there' and 'now' and 'yesterday'.[39]

On Newton's ontology of existence, then, infinite space exists in so far as it depends on Divine nature *qua* existent. As such it is not a substance, given that substances are individuals which have independent existence. But this in itself does not prevent space's being a substance in some other sense. In fact Newton makes the following claim:

> ... but it is usually believed that these spaces [those within and without the finite world] are nothing (*sunt nihil*). Yet they are true spaces. Although space may be empty of body, yet it is not itself void. And there is something because spaces are there, although nothing else besides that.[40]

Here again Newton refers to the nature of space as an existing thing, and has in mind a decided list of properties which he considers belong properly to its nature: it has inherent parts that constitute its structure; in all directions it possesses unlimited extension; it is actually infinite; independently of things external to it, its inherent parts maintain the same eternal and immutable order; all true "positions, distances and local motions" have reference to these parts alone; and lastly, space is passively receptive to being occupied, and can offer no resistance to motion.[41] It is clear that Newton considers space to be an individual in the sense of being the proper subject of certain sorts of predications. "There is no idea of nothing, nor has nothing any properties, but we all have a very clear idea of extension, abstracting the affections and properties of a body so that there remains only space ...."[42] This passage, and the previous ones quoted, leave no doubt that Newton beieves space to be a positive individual embodying real properties which derive from its actual extension in three dimensions. As such, it is not a form of non-being, but an individual fully actual at all time, and endowed with a rich inherent structure. This view of the nature of space is probably what Newton has in mind when he observes that space "is more thing-like than an accident, and closely approaches the nature of substance."[43] In other words, space is a substance in the minimal sense; it is that of which specific things can be said.

So far Newton has denied that space is either a fully-fledged substance or a real attribute of substance. To say this, however, is not to say what space is positively. This brings us to the second part of Newton's claim, namely, that the extension of space can be distinguished from corporeal extension, despite the Cartesian arguments to the contrary. We begin by asking: what response can Newton make to the view that only the corporeal can be extended and dimensional? First, it is well to be clear about the kinds of responses which he rejects.

He does not argue in the manner of Henry More. For the latter, anything that can be said to exist must be extended, God included. Taking the category of extended being as the *summum genus*, More argues that its subaltern *genera* are corporeal and incorporeal substance, individuals which have only the property of 'being extended' in common.[44] Given this framework, he argues that we can speak of dimensions which, though real, are not modes of corporeal substance as Descartes demands. For More the differentia for the subaltern genera is this: corporeal dimensions are discerpible, i.e. they possess parts capable of being separate; whereas incorporeal dimensions are indiscerpible, i.e. their parts are indivisible in the sense that they cannot be separated from one another.[45] And the ground for the inseparability of the 'parts' of incorporeal dimensions is that they are modes of an attribute which presupposes the unity of incorporeal subtance.[46] Apart from the fact that More's position begs the question against Descartes, Newton cannot accept it, because he denies that space's extension is a real attribute of incorporeal substance.

But neither does he attempt to break the connection between dimensionality and corporeality in the manner of Francesco Patrizi. The latter argues that space is prior to, and separate from, body, whose primary attribute is resistance. Thus in Patrizi's view, space is essentially tri-dimensional; whereas tri-dimensionality is assigned to the corporeal only accidentally.[47] But Newton cannot accept that tri-dimensionality only pertains to the corporeal accidentally, i.e. as an accident of corporeal nature. After all, integral to his position is the claim that the dimensions which we ascribe to corporeal objects are grounded in the presence of the volumetric figures which inhere in the structure of space itself. Since the dimensions of these figures are a part of the total nature of space, any distinction between essential and accidental tri-dimensionality does not apply.

Nor does Newton argue like Pascal that the "essential difference between empty space and body, which has length, breadth and depth, is that one is immobile and the other is mobile, and that one can receive within itself a body which penetrates its dimensions, whereas the other cannot." Pascal's argument turns on the assumption that the "impenetrability of dimensions is to be understood only of the dimensions of two material bodies,"[48] but not of the nature of dimensions generically understood. But this argument merely *assumes* that we have an idea of dimensionality other than that of material things. It fails, therefore, to counter the Cartesian claim that dimensionality is inextricably connected with corporeality.

Each of these arguments is based on one or more of the following distinctions: that there is a difference between corporeal and incorporeal substance; that there is a generic notion of dimensionality which can be differentiated into sorts by the notions of divisibility and indivisibility, or by those of penetrability and impenetrability; and, lastly, that tri-dimensionality belongs to the corporeal derivatively but not essentially. But each argument begs the question, in that the respective terms employed to distinguish corporeal from incorporeal dimensions already presupposes that there are independent grounds for establishing the separate existence of these natures.

Although Newton employs these distinctions, or their near cognates, when speaking of dimensions, he never asks directly whether the dimensions of infinite space are corporeal or incorporeal, anymore than he asks whether space itself is a corporeal or an incorporeal substance. It is clear, then, that he rejects the standard ontological moves advanced by those wishing to counter arguments to the effect that dimensionality pertains only to the corporeal. And in so far as he does, he appears to avoid their difficulties. But in the end, does he not succeed in cutting the ontological ground from underneath himself? After all Newton wants to distinguish the "amplitude of presence" of corporeal things from the infinite extension of space itself. However, when he does mention indivisibility, explicitly, what he says is that "spaces themselves are not actually divisible (*spatia non esse actu divisibilia*)" whereas corporeal things are "extended and consist of divisible parts."[49] All this says, however, is that unlike corporeal things, space's (i.e. places) do not possess 'parts' which in reality can be separated from one another. What else could be meant by the phrase "not actually divisible?" Moreover, "spaces themselves" are the macro-figures which, though in themselves unsplitable, are of course conceptually divisible; so no reference is made to the ultimate indivisibility of space itself. As far as it goes, then, there is nothing in the distinction Newton draws with which More would disagree. Moreover, it satisfies More's sense of spatial indivisibility, i.e. that which lacks discerpible parts. But More attempts to ground his position by arguing that space is itself an attribute of incorporeal substance. It is this sort of claim, however, that Newton denies. How, then, can he provide ontological support for the distinction he draws?

Newton employs another argument which he thinks pertains to the infinite extension of space. It turns on distinguishing what can be conceived by the imagination and what can be grasped by the understanding:

If anyone now objects that we cannot imagine that there is infinite extension, I agree. But at the same time, I contend that we can understand it. We can imagine a greater extension and then a greater one, but we understand [*intelligimus*] that there exists a greater extension than any we can imagine. And here, straightway, the faculty of understanding is clearly distinguished from imagination.[50]

Newton appeals here to epistemological distinctions prevalent in the seventeenth century as in earlier thought. At the direct sensory level, the mind only apprehends particulars via determinate modes of the properties which they instantiate. The imagination, on the other hand, though it can construct composites of considerable complexity which surpass immediate experience, cannot generate a clear mental picture of anything which transcends its powers. Although Newton makes no explicit mention of the senses in this context, it is clear that he would deny that they provide a sufficient basis for apprehending the infinity of space's extension. Similarly, the powers of the imagination are unable to construct a complete representation of its vast and limitless nature. Only the understanding is capable of making the infinite nature of space its proper object, where the term 'object' connotes that which is appropriate to that way of knowing. So while imagination in an act of mental picturing cannot represent the vast stretches of infinite space, the understanding can grasp the fact that space is an infinite and individual nature, even though it lacks a complete notion of what it is.[51]

Notice that Newton distinguishes the faculties of imagination and understanding, including what is appropriate to their respective ways of knowing. If the nature itself of infinitely extended space is grasped only by the understanding, Newton is committed to the belief that in so far as space is known, this cannot be on the basis of the senses or via the powers of the imagination. And this commitment involves another: that extended space is an "eternal, infinite and uncreated" entity which, by its very nature, transcends the powers of the senses and the imagination. In other words, extension is an actually infinite individual; as such it is an object of the intellect alone.

The thrust of Newton's argument can be developed in terms of a distinction which, if not explicit in his position, is at least consistent with it. It is the distinction between that which perfectly satisfies a given genus, and that which lacks in some determinate ways with respect to that genus. When this distinction is applied to the problem of distinguishing infinite extension from corporeal extension the argument is as follows: with respect to the genus extension, infinite space is perfect: it is unlimited in terms of tri-dimensionality. Infinity thus completes the perfection of extension, since "infinity of extension is so far perfect as that which is extended."[52] Since they are bounded by determinate shapes and sizes, corporeally extended things are less perfect of the kind. Accordingly, they lack with respect to the genus extension because of the particularity of their individual natures. But the infinite extension of space is not grasped by the imagination; rather, the fact that its nature perfectly satisfies the genus extension is established by the understanding. Consequently, the understanding grasps infinite extension as an eternal and uncreated nature, a capacity which transcends the power of either sense or imagination. Thus, there are grounds for distinguishing space's infinite dimensions from those of corporeal dimensions, since space is an eternal and uncreated nature and an appropriate object of knowledge for the understanding alone.

I shall return presently to the question whether Newton can consistently accept the epistemology involved in this development of the argument. But first we must ask: is being infinitely extended sufficient to distinguish space, as an object of the understanding, from corporeal dimension, as an object of the imagination? After all, being infinite, does not in itself distinguish natures; rather, to say of a nature that it is infinite is to say something about that nature. This, in fact, is a view which Newton argues for explicitly. Infinity, he tells us, is not a real attribute which 'adds' something to the intrinsic nature of anything, but rather a second-level characterization of that nature's perfections or imperfections. So while it is not itself a perfection, infinity nevertheless 'completes' the perfections or imperfections of a given nature, where to be a nature is to specify a genus, kind or sort. Infinity is thus a perfection only "in so far as it is ascribed to perfections."[53]

What, then, is the perfection embodied in infinite extension itself, considered generically, that the understanding is able to grasp, and which distinguishes it from corporeal extension? It can only be that the mind apprehends a positive connection between extension's being at once an infinite *and* indivisible nature; whereas, in the case of corporeal extension, the mind perceives the connection between its finiteness and its divisibility. It seems that Newton implicitly accepts this reasoning in that he claims extended space is *in fact* infinite, and, moreover, that the mind can have a positive conception of its infinity. And it is infinite because it is without boundaries; hence, it is impossible that anything can be said to exist beyond it. This view finds support in Newton's explicit denial of Descartes' claim that generic extension is not the fact infinite, but rather an indefinitely divisible entity.[54] And it finds further support in Descartes' claim (which Newton also rejects) that extension is only conceived indeterminately by the imagination, and not positively affirmed by the understanding.[55] But Spinoza also denies the

Cartesian claim for reaspons which appear similar to Newton's. And, like Newton, his arguments employ the distinction between the intellect or understanding as it pertains to infinite extension generically considered, and the imagination as it relates to corporeal extension. It will be instructive to consider how Spinoza develops the argument, since it will enable us to see, by contrast, the full implications of Newton's position on the unity of space.

With respect to the imagination, Spinoza argues, extension is finite, divisible and composed of parts; but with respect to "the intellect . . . we find it to be infinite, one and indivisible."[56] Unlike Newton, however, Spinoza explicitly bases the distinction on fundamental doctrines. In terms of Spinoza's conceptual framework they are the following. First, he holds that there is uniquely one substance, which exists in itself and is conceived through itself. That is, in Spinoza's view both the grounds for substantial existence and the basis for conceiving it are intrinsic to that nature alone.[57] Secondly, he argues that substantial reality is expressed through each of its infinite attributes which are indistinguishable, *in re*, from the unity of substance itself.[58] Lastly, he holds that there are levels of knowledge, and that extension, as an object of the imagination, can be only known confusedly and incompletely.[59]

By means of these doctrines, Spinoza argues that infinity and indivisibility are unified in the essential nature of substance. Since substance is that which exists in itself and not in another, it grounds its own existence. Anything whose ground for existing is itself, is a necessary existent. But a necessary existent is that which is uncreated and eternal in nature. And to be uncreated and eternal in nature, is to exist unqualifiedly and independently of all else. Anything which thus exists is substance as Spinoza conceives it; namely, a unique and basic individual whose existing *in se* relates intrinsically to its conception *per se.*[60]

But if substance exists unqualifiedly, such that apart from it there is no separate reality, it is an absolutely infinite being. According to Spinoza, this is to say that substance is an infinitude of infinite attributes, each an expession of its essential nature.[61] Were its nature otherwise, substance would lack the perfections embodied in an infinity of infinite attributes, and in Spinoza's sense would fail to be absolutely infinite. Not only is substance an infinitude of infinite attributes, but each attribute is unlimited in terms of its kind and perfection.[62] So not only does the mind conceive that substance is an infinitude of infinite attributes, but it apprehends that their infinite natures imply no reciprocal limitation.

Spinoza bases this last claim on a definite view of how substance and attribute relate. He denies that their relation is to be construed on the model of properties said 'to attach' to things. On this view, substance is conceived as an aggregation of 'properties', each less than the whole, and thus mutually limiting in their expression of its complete nature. But for Spinoza, the infinity of substance is not the totality of its attributes; for anything which is a totality, necessarily involves composition and thus the mutual limitation of its 'parts'. Rather, each infinite attribute is identical *in re* with substance itself and, as conceived through itself, "each expresses the reality or being of substance."[63] The finite mind, however, is capable of apprehending it only through the content of the infinite attributes of thought and extension. Substance so conceived is identified by Spinoza with God or nature.

It is in terms of this analysis of the unity of substance and attribute that Spinoza conceives the relation between the infinity and indivisibility of extension. That is to say, apprehended through its intrinsic nature, extension is an actually infinite attribute which pertains to the being and unity of substance itself.[64] When so conceived, its nature is not grasped *partes extra partes*, as a quantity is grasped by imagination. Rather, it is conceived in the manner proper to the understanding, as an infinite and indivisible nature appropriate to the absolute infinity of substance.[65] As such, the true unity of extension is apprehended, and not merely the unity of aggregation appropriate to the way of knowing of the imagination.

Notice that Spinoza provides an ontological basis for the *epistemological* distinction between objects of imagination and understanding. With respect to imagination, extension is represented confusedly and incompletely, whereas the understanding grasps its nature clearly and distinctly.[66] In both cases, the *same* nature is apprehended. In the latter case infinite extension is apprehended in terms of the unified being of substance on which it depends ontologically. As such, its infinity and indivisibility are seen to be connected. And since it is the same nature grasped by both faculties, ultimately the same reality grounds both infinite and corporeal extension.[67] As it appears to imagination, extension is finite, divisible and composed of parts. This, however, is merely a confused perception of its infinite and indivisible nature.

As I suggest earlier, Newton seems implicitly to accept that the understanding grasps a connection between extension's being actually infinite and its being indivisible. In Spinoza's ontology, this feature of extension is grounded by its being a real attribute which pertains to the true unity of Divine substance itself. But in Newton's view, the extension of space is not a real attribute of God's essential nature, but rather an affection which pertains to the actuality of Divine existence.[68] For the same reason, extension is not a substance, as Newton readily admits, in the sense of being a basic individual which grounds its own existence. But if this is the case, there is nothing in the gross ontology of extension itself, as Newton conceives it, which provides a basis for linking infinity and indivisibility. To be sure, he claims that extensions in space are indivisible. They are indivisible, however, only in the sense that they cannot actually be divided or separated from themselves or from space. But there is nothing in the modal characterization 'cannot actually be – separated from itself' which shows that extension in itself is also to be characterized as infinite. Equally, to say that extension is an infinite affection of Divine existence itself is not to show that it is a determinate and complete individual; i.e. that it is a being which is intrinsically one and indivisible in Spinoza's sense of these terms.

In other words, there is nothing in Newton's conception of the overall extension of space which, in itself, mutually implicates the notions of indivisibility and actual infinity. For Newton nowhere positively identifies the sort of being that space itself satisfies in so far as it manifests these characterizations. It is precisely this relationship, however, that Spinoza's ontology displays. For it is only through apprehending the *unity* of Divine substance itself, according to Spinoza, that the understanding is able to grasp generic extension as at once an indivisible and infinite concretum.[69] And if infinite extension is to be distinguished from corporeal extension, as Newton wants, he must show it to be conceptually indivisible in accordance with its intrinsic nature. Otherwise he cannot successfully reject Descartes' claim that extension pertains only to the corporeal and divisible, while asserting simultaneously that extension itself is actually infinite.

Newton is in a crux. If he accepts Spinoza's view of the infinity and indivisibility of extension, the characteristics of the latter's ontology are forced upon him. He has to accept, that is, an ontology in which God's attributes are strictly one and the same with Divine essence, as well as identical among themselves.[70] In this case, the infinite extension of space becomes an intrinsic attribute of God's defining nature, and Newton is moved towards a position which he would regard as a form of pantheism. Furthermore, an acceptance of an ontology akin to Spinoza's means that Newton forfeits the foundations of his theory of absolute space. In other words, he forfeits the doctrine that space is an extrinsic denomination of Divine presence itself, where to speak of God's "amplitude of presence" is just to say that he exists infinitely with respect to the uncreated extension of space.[71] But in so far as he maintains his characteristic ontology, Newton lacks a vehicle for displaying the essential unity of infinity and indivisibility necessary to the generic nature of extension itself. It is clear, then, that a close relationship exists between Newton's denying that infinite extension fits either the ontology of substance or its essential attribute, and a failure to justify a distinction between the unity of infinite extension in itself and the aggregational nature of corporeal extension.

This conclusion should not surprise us. An 'Eleatic' conception of the unity of being (an undifferentiated whole) is incompatible with Newton's ontology of Divine existence, and with his conception of generic extension's inherent structure. Its unity must lie elsewhere. It is on this question that Newton would certainly mobilize his theory of spatial minima. If Newton's space cannot be everywhere and indivisibly the same, its ultimate ground of unity rests with the conceptual impossibility that its inherent and partless minima be divided by the mind; for "ye nature of union (being but a modall ens) is to depend on its pts wch are absolute entities". (*Questiones*, f. 88^{r}). On this view, infinite space is a unity of infinitely many indivisible yet extended minima, which together compose an extended whole, itself differentiated and thus conceptually divisible. Moreover, throughout the eternal reaches of infinite space, minima are everywhere and immutable interlocked: it is therefore absurd to suppose that spatial unity is threatened by forms of change. It is clear that corporeal extension satisfies none of these criteria, since what is corporeal in nature is both conceptually *and* physically divisible. (But not Newton's macro-figures. When materially delineated, they are conceptually, though not actually divisible). And even though on atomistic principles, corporeal entities are made up of indivisible parts, their minima are in principle separable from one another spatially. But this conception lacks sense when applied to the minima of infinitely extended space.

Newton's conception of extension's unity is not without its problems. Since he cannot accept Spinoza's account of it, how is Newton to relate its infinity and indivisibility? Do space's inherent figures not oppose its infinity? Since Newton claims that the extension of space is actually infinite, should it not exist (like the Eleatic one) without internal differences to limit it, just as it is impossible that anything can limit it by existing beyond it? But space can be conceptually differentiated in virtue of the differences of its figures. In response, Newton can mobilize his conception that infinity is a second-level characterization of existing natures. Since extension and its indivisible figures are mutually compatible in nature, one can be characterized as infinite in magnitude, the other as infinite in number. But granted that compossible natures can be said to be infinite according to their kinds, the relation between space's infinity and indivisibility is not solved. The difficulty remains that the intrinsic *finiteness* of extension's figures, though in themselves said to be indivisible, is irreconcilable with Newton's conception of its actual infinity – if that is construed as what lacks all limitation.

A problem of unity arises in another way. Newton claims that infinite extension is itself a positive reality grasped as a whole by the understanding. It is an eternal, infinite and uncreated *aliquid*, whose nature the mind apprehends in itself prior to any of its specific determinations. This presupposes a unity intrinsic to the individual nature of extension which differs from the composite unity of its inherent structure. Moreover, the latter, taken by itself, is such that it can be successively pictured by the imagination, and does not depend on the powers of the understanding. While there is no intrinsic incompatibility between these levels of unity, unlike Spinoza, Newton does not give a direct account of their epistemological connections.

This brings us back to an earlier question: can Newton accept the epistemological intuition which underlies the distinction between objects of the understanding and the imagination? It is this: in order to know that something is imperfect with respect to a kind, it is necessary to have a cognitively prior grasp of the perfection of that kind. For example, we would not teach someone to play perfectly in tune, by playing out of tune, and say "but not like that." Rather, the notion of "playing out of tune" involves a prior and positive conception of what it is to play perfectly. Similarly, in accordance with this principle, knowing that something imperfectly manifests the genus extension, depends on a cognitively prior grasp of the positive perfection of that genus, i.e., of infinite extension.

But a word of caution is necessary. Spinoza holds that in the experience of an individual, the understanding can apprehend the sort or kind which that individual concretely instantiates. In other words, he accepts the basically Aristotelian view that the mind can grasp the general in particular cases.[73] This does not mean, however, that a cognitively prior conception is first in the order of experience. After all, on the basis of one and the same experience, the mind is capable of producing either confused and incomplete ideas of an individual or of grasping the essential in its nature. But in so far as there are grounds for saying that an individual is known imperfectly, there is presupposed an implicitly general conception of what it is essentially. Thus, on the basis of one and the same kind of experience the understanding, when applied attentively, not the imagination, grasps the individual's essential characteristics as a general sort or kind. And it then makes *explicit* those general characteristics of the nature that are embodied in the genus, sort or kind which the individual instantiates.

We are now in a position to appreciate some important connections in

Spinoza's view of the epistemology and ontology of extension. Knowing extension by means of either the imagination or the understanding does not mean that the mind apprehends isolated bits of knowledge. On the contrary, the two levels are interconnected, since the former presupposes the latter. However, in the order of being or nature, extension is prior to its conceptualization, but in the order of conceptualization both levels of knowledge are prior to extension itself.

Much that Newton says about extension is consistent with his acceptance of these epistemological and ontological interconnections. As we have seen, he holds that spatial extension is an infinite, eternal and uncreated nature, capable of being grasped by the understanding alone. After all he states that while we can repetitiously imagine successive bits of finite extension, yet "we understand that there exists a greater extension than any we can imagine."[74] Second, he insists that the understanding has an "exceptionally clear idea of extension" as a defining characteristic of space itself, for it is "real and intelligible in and of itself."[75] Lastly, he argues that relative conceptions of extended space are merely "extrinsic denominations," or conceptions consistent with imagination alone: as such, they are incapable of grasping the real nature of extended space as it exists *per se.*[76]

At this point, an unmistakable gap appears between Newton's explicit empiricist epistemology and his view that the understanding alone apprehends the infinite nature of extended space. In so far as he accepts that there are natures, and especially that spatial extension is an uncreated nature, Newton is a Platonist in spirit. And it is no surprise that he insisted that only the understanding can form a conception of the actually infinite extension of space. However, Newton gives no specific account of how the mind intuits an extra-mental nature such as spatial extension. At best he speaks explicitly of the mind's capacity for 'abstracting' generic extension from the affections and properties of corporeal natures.[77] All this says, however, is that the nature embodied in generic 'extension' can be expressed as a general term or a linguistic universal constructed on the basis of perceived resemblances among individual bits of extension. In other words, the *idea* of the infinity of generic extension is based on the imagination's ability endlessly to multiply finite quantities of extension. On this view, there is an idea of an infinity of quantity, but no positive idea of actually infinite extension.[78] The latter is quantity than which there could be no greater; this conception, however, is contradictory to the notion of quantity as here supposed.

So extension in itself cannot be an actual or complete infinity; nor can the mind have any idea of it, since that entails the absurdity of having a positive idea of a never-ending progression. There is no doubt, however, that Newton thinks of space as a quantity existing *partes extra partes,*[79] whose idea is a mental picture. But this is consistent with holding that spatial extension is empirically determinable on the basis of sensory experience.

But this explicit epistemology is hardly consistent with his claim that extension is a complete and actually infinite nature, which exists extramentally. Nor with his further claim that, though it may be aided by the imagination as its instrument, in the end the nature of extension is alone apprehended by the power of the understanding. Here in *De Gravitatione* is the first indication of an unresolved tension that remains in Newton's thought. On the one hand, he tends to assimilate the content of the intellect to that of the sensory, and characteristically holds that knowledge of general truths at the perceptual level is *inductively* grounded in the deliverances of perceptual experience.[80] But, on the other, he insists that knowledge of the actual infinity of extended space is an affair of the understanding alone, in so far as that capacity is able to grasp the uncreated nature of extension itself. Unfortunately, the tension between Newton's sensory epistemology and his ontology of extension is nowhere put to rest. In fact, in the Scholium on space and time of the *Principia* (1687), Newton tells us that though in common affairs sensible measures of things suffice, when they are considered truly and "philosophically, we ought to abstract from our senses."[81] Ironically enough, this can mean either of two things: that we isolate the confusions of the senses, or that we concentrate on the general characteristics of space independently of the process whereby they are 'abstracted' from the particulars of sense experience.

III. EXTENSION AND CORPOREAL NATURE: NEWTON ON THE CREATION OF MATTER

I turn now to Newton's account of the creation of matter. In the first instance, it is well to note that he gives no account as to why infinite extension contains an infinitude of different kinds of volumetric figures. His view seems to rest on the unexamined assumption that in uncreated natures, such as infinite extension, all consistent possibilities are actual. Thus the positive tridimensionality of space is not merely consistent with the possibility of its containing an infinitude of geometrical solids, rather it actually does contain them. In other words, if space were totally unstructured, ultimately it would be unintelligible to the human mind, since it would lack the geometrical determinateness appropriate to the form of being that it is. However, given

Newton's view of the structure of space, he advances the following claim: " . . . the material delineation of any figure is not a new production of that figure as regards space, but only a corporeal representation of it such that what was first insensible in space now appears to exist to the senses."[82] Newton illustrates his meaning by a useful if not exact analogy. If colors (dyes) are added to otherwise clear water, distinguishable shapes appear, the boundaries of which exist before the colors make them visible. Likewise, Newton argues, the inherent figures in space can be made visible under certain conditions.[83] But what are these conditions, and how are they caused?

It is well to remind ourselves of Newton's motivation for advancing an account of how God could have created matter. First, he wants to show that matter is utterly dependent on Divine will in order to counter what he sees as the atheistical consequences of Descartes' conception of corporeal extension; that is, the conception that matter exists not by Divine will but of necessity. Second, he wants to avoid the substratum view of substance construed as the ultimate and unknowable bearer of property instantiation. Though he gives no particular account of it, Newton seems to suppose that in God all possibilities are present. Indeed, such an assumption would seem to be behind the following thesis: we cannot say that God could create matter in one way only, since there are "several ways by which different beings similar to bodies could be produced." Thus, we cannot deny Divine power the ability to create "beings similar to bodies which display all their actions and exhibit all their phenomena, and yet in their essential and metaphysical constitution are not bodies." So while we may never know the ultimate nature of matter, we can nevertheless advance a coherent account of its creation which is "within the power of God, so that we can hardly say that it is not body."[84] Thus, Newton speculates that God can so condition particular regions in infinite extension that certain of their inherent figures are able to cause the properties which are attributed to bodies through sensory experience. Hence what was previously an invisible figure in space is now able to produce the perceptible properties of matter itself.

Fortunately, Newton provides further details as to how he thinks that God could have brought matter into being. If through the exercise of thought and will alone God can move bodies, he can also, in virtue of omnipotence, "cause the boundary of some space above the earth, such as a mountain or any other body, to become impassable to bodies and thus stop or reflect light and all impinging things." Consequently, if so conditioned, that particular space will be "tangible on account of its impenetrability, and visible, opaque and colored on account of the reflection of light, and will resonate when struck because the adjacent air will be moved by the blow." In general, it is conceivable that God conditions a number of these inherent figures, such that they are mutually impenetrable, and capable of being transferred across space while preserved as definite quantities of impenetrability. Thus considered, each one "would have shape, be tangible and mobile, and be capable of reflecting and being reflected, and no less constitute a part of the structure of things than any other corpuscle . . . ."[85] And though these 'beings' existing in space are different in nature from solid bodies, it is clear that they are capable of producing similar sensations and perceptions. For given their causal properties, there is no ground for supposing that they are unable to affect perception via the senses, and conversely be affected by our actions. In the end, they are "nothing more than the product of the Divine mind realized in a definite quantity of space;"[86] as such they exist in space and are independent of the perceptions which they cause.

Accordingly, Newton's conception of how God creates matter is as follows. Given that Divine Power and infinitely extended space co-exist, God can 'pick out' adjacent figures in space and endow them with a causal capacity to interact mutually and to excite perceptions in the mind. When so endowed each of these figures 'limits' the causal actualization of other figures co-existing in that respective region of space. For were other figures in a region actualized, more than one material being would occupy the same place, which Newton defines as that "which something fills evenly." From the impossibility of more than one thing evenly filling the same place, Newton derives the mutual impenetrability of the materially delineated figures. Furthermore, he holds that it is consistent with his hypothesis "that that impenetrability is not always conserved in the same part of space, but can be transferred hither and thither according to certain laws, yet so that the quantity and shape of that impenetrable space are not changed."[87] And the "quantity and shape" do not change because each space made impenetrable by God, possesses similar inherent figures, e.g., if the materially delineated figure is spherical, all figures in space successively endowed with the capacity to cause a representation of a material sphere are spherical prior to being conditioned by Divine power. Thus our perceptions of visible material objects are ultimately grounded in God's plan to cause figures inherent in space to produce experiences which are similar to those we would have if the solid corpuscles of the mechanical philosophy in fact existed. So whether these beings are bodies or like bodies, they are able to cause sensations and ground sense-perceptions. And we are able to judge this because "whatever reality we attribute to bodies arises from their phenomena and sensible qualities."[88]

But what is the ontological nature of these beings which produce sensations and provide the basis for perceiving material phenomena? And how does God accomplish these effects via the mediation of space? Let us consider the second question first. Newton distinguishes three things: Divine will, the effects of Divine will, and space itself which receives these effects. "But I do not see why God himself does not directly inform space with bodies; so long as we distinguish the formal reason of bodies from the act of Divine will."[89] In order to explain how God acts on space Newton invokes an analogy. As we are aware of the fact that we can move our bodily parts as well as other bodily things by the power of the will alone, so we can understand that the immensity of Divine power is capable of that, and also of making particular figures impenetrable.[90] Nevertheless, Newton admits that while we can understand the claim that God is in fact able to do this, we cannot conceive the manner in which he so acts. However, if we could understand how the will moves bodies, and given that there is an "analogy between the Divine faculties and our own," we would be able to grasp how Divine omnipotence creates matter. Thus Newton concludes that difficulties in the "conception may at length be reduced to that."[91]

It is ironic, however, that he should attempt to explain how God acts on space within a strict voluntarist framework. Indeed, voluntarist theologies invoke Divine omnipotence as the principal attribute and ground of creation, and claim that we have no conception of how or why God creates. For it is characteristic of voluntarism to hold that the archetypes of creation, as well as God's reason for enacting them, are beyond human comprehension.[92] Nevertheless, the central role that voluntarism plays in Newton's thought should not be underestimated. It is essential to his conception of both Divine and human nature. God is a conscious agent, and infinitely capable of purposive action in the realization of his intentions. Human action is also intentional, and imperfectly mirrors God's volitional nature. But Newton's voluntarism also involves a definite view of the nature of causation. Only agents which in themselves are capable of purposive action are able to bring something about in the act of realizing their intentions. Hence change, activity and motion ultimately result from the conscious behavior of intentional agents. And in so far as space itself is endowed with the capacity to affect our sensory organs, this results from the created effects of Divine will. In short, efficient causation is inseparably connected with intentional action that leads to the realization of conscious ends.[93] It is clear, therefore, that the passivity of extended space cannot ground change in itself or in things other than itself. Consequently, the active properties of corporeal nature are superadded to space via the mediation of Divine will.

As to the first question – the ontological nature of the beings that God creates in space – Newton raises some interesting issues. In fact, he presents two separate but related descriptions of the matter's ontology which illustrate different aspects of his theory. On the one hand, Newton says that God enows certain figures in space with active powers to affect other things and to excite perceptions in the mind through the deliverances of the sensory organs.[94] On this view, the essential nature of matter is its power to affect or its capacity to be affected via the mediation of the place that the power occupies. Two important bits of evidence show this aspect of Newton's conception of matter. First, his insistance that bodies are substances only in so far as they are capable of moving, of offering resistance, and of exciting perceptions.[95] It is these features that distinguish bodies from the reality of extended space, which in itself cannot ground change or motion. Second, given that there is a likeness between the nature of Divine action and the character of its products, the result of God's acting on "determinate quantities of extension" is to endow them with an individual efficacy which has Divine will as its immediate ground.[96] Nevertheless we must distinguish God's will from the ontology of bodies which are its effect. For, according to Newton, it is contradictory that body "should be the act of willing or anything other than the effect that that act produces in space."[97] It is on the basis of these commitments that Newton claims that space is ontologically autonomous and really distinct from matter conceived as a set of active powers.

On the other hand, Newton uses a different set of concepts to characterize the features of matter as they relate to space. In order to provide further classification of the ontological character of the conditioned regions of space, he observes:

> That for the existence of these beings it is not necessary that we suppose some unintelligible substance to exist in which as subject there may be an inherent substantial form: extension and the act of Divine will are sufficient. Extension, in which the form of body comes about and is conserved by Divine will, takes the place of the substantial subject; and that effect of the Divine will is the form or formal reason of the body denominating all dimensions of space in which the body is to be produced.[98]

There are a number of important points here. Most striking is the use of the conception of formal causation. Newton conceives the effect of God's acting on a determinate figure in space as imparting to that space the formal cause of matter. In other words, God informs particular figures in space with that which makes matter what it is; namely, that which is capable of affecting things, of moving and of affecting the senses. In so doing, he actualizes its formal nature which prior to that act exists in the Divine mind *qua* idea or archetype. Thus instead of the conception of substances as distinct individuals comprising matter and form in the composite, Newton conceives infinite extension itself as receiving the form of matter through the direct action of Divine will. So conceived, uncreated and extended space can be viewed as a 'receptacle' within which God himself imposes the formal nature of material things. Hence contrary to his tentative entertainment of the analogy, Newton expressly denies that the relationship between extension and its impressed form is akin to that between *materia prima* and substantial forms. This analogy invokes the picture – at least in terms of the scholasticized Aristotelianism which Newton rejects – of an undifferentiated, qualityless substratum which is informed with some determinate set of characteristics. But this conception is unintelligible to Newton. In fact he says that "extension (since it is a definite sort of quantity) has more reality than *materia prima*."[99] Its individual nature has positive and determinate features *per se*, unlike what Newton takes the scholastic conception of prime matter to be: that which in itself is nondetermined and hence a form of non-being. Thus in the spirit of Plato's *Timaeus* Newton conceives the reality of structured space as an eternal and unchanging matrix into which Divine will impresses the form of matter. And consequently he rejects what is in fact a scholastic version of Aristotle's doctrine of first matter as that undifferentiated thing out of which informed individuals are said to come to be.[100] Infinite space is thus the ground in which the genesis of matter takes place. And when God actualizes the essence or formal nature of matter it has the capacity, *qua* created being, to ground those determinate features which are ascribed to corporeal phenomena in virtue of the representations that they cause the mind to have.

It is clear, then, that in Newton's view matter is the actualized embodiment of Divine power in determinate parts of infinitely extended space. The ontology of *De Gravitatione* can now be summarized as follows: there exists God's active will along with created finite wills; and coexistent with Divine presence, as a co-principle of material creation, is uncreated extension whose inherent parts are endowable by God with active powers. Thus God, created wills and matter, *qua* powers, are active entities, all of which are distinguishable from infinite extension. But it is extension, when conjoined with active powers, that provides the causal basis for the mind's representation of both the primary qualities of matter, such as impenetrability, and sensations of colors, sounds and touch.

Unfortunately, neither of Newton's accounts of matter answers an important question: how are the active powers with which God superveniently endows space related to the figures in which they exist? This is in effect the problem of the unity and individuation of a determinate 'piece' of matter. It is clear that Newton implicitly identifies material phenomena through their operations. And clear also that he considers the notion of impenetrability as derivative epistemologically from the more basic notion of things resisting one another via their intrinsic powers. But what are those powers powers of? There is clearly no intrinsic connection between them and the extension of the macro-figures as such. Nor can Newton claim that powers are reducible to a micro-structure, since it is the essence of his position to deny that the properties of perceptible matter are grounded in an infrastructure of the kind posited by the mechanical philosophers. Unless he can give a coherent account (which he recognizes but does not attempt) of how God 'picks out' and endows a particular figure with individual powers, Newton cannot answer the difficulty within the ontological framework which he allows. The ghost of rejected substance returns to haunt him.

At this point, Newton's account of the ontology of motion requires further comment. If the metaphysical foundation of the corporeal is in fact figured extensions endowed with the requisite causal powers, in order that perceptible motion can be accounted for these powers must be moveable. But how is this possible? And how is one and the same quantity of extension identical over time? The answer to these questions follows directly from Newton's voluntarist theology. God identifies a range of figures of similar sizes and shapes (e.g., spherical), and endows each in turn with the same causal powers. In other words, he maintains and preserves the same active conditions in each inherent figure of the same determinate sort. This means that the mind's notion of the permanence and moveability of matter is grounded in the fact that God continuously acts on space in the same immutable way. This claim is an expression of the standard theological view that God always acts in the simplest manner so as to achieve his ends; in this case the preservation of similar effects in separate regions of space. Thus, instead of the unity of matter and form, as in the Aristotelian tradition, Newton posits the unity of the effects of Divine will as they inform uncreated space. And since these effects 'materially delineate' figures of the same determinate sort, *similar* sorts of geometrical figures are 'materially' identifiable over time.

This does not mean, however, that matter is itself identified with any region in space. For "spaces are not very bodies themselves but are only the places in which bodies exist and move."[101] Nor does it mean that separate geometrical regions are identifiable with each other in virtue of God's materially delineating them. For in terms of Newton's conception of the structure of space, quantities of extension (parts in space) cannot be numerically identical with each other. In other words, God's informing region x with b at t, and region y with b at t_1, does not entail the identity of x with y. Rather, the notion of something's being the same matter is explicated by the fact that matter "borrows the denomination of numerical body from its form. For so I suppose that any form can be transferred through any space, and the same body is everywhere denominated."[102] Thus the identity of the same quantity of matter is grounded in the fact that God produces and instantiates the same formal nature in one particular figure or in separate but similar figures at different times. And it is this formal nature which embodied the criteria of identification of the sort or kind that matter is. Here Newton turns the Aristotelian principle of individuation on its head. Instead of matter individuating a particular manifestion of a form, it is form itself that qualitatively differentiates certain figures in space. The result is the production of actual material beings where previously the form of matter was non-instantiated. Furthermore, continuity of motion is not a result of God's continually *recreating* the same conditions in successive regions of space; but rather of his maintaining, through the laws of his *potentia ordinata*, the same formal reality in different parts of space through successive times. Continuity of motion is therefore real rather than apparent.

It is clear, then, that Newton reject Descartes' claim that being extended is both necessary and sufficient for being a corporeal nature. Given the ontological priority of infinite extension it is necessary, in Newton's view, that any corporeal nature be identified *via* particular figures inherent in the structure of space. That in itself, however, does not define it sufficiently. For it is tangible, moveable, capable of affecting the senses and of offering resistance. According to Newton, each of these features is defined primarily in relation to the sensory modalities, and characterizes a nature which any mind can grasp.[103] Moreover, were any of these sortal characters removed from corporeal being, the result would be "to reduce body into empty space."[104]

But a large question remains. What are the grounds of individuation of the figures inherent in infinite extension? It is clear that the grounds for different instantiations of one and the same form (i.e., matter) are the diverse figures inherent in space itself (along with determinate times). So in Newton's view, 'this' particular instantiation of the form of matter has as its referent 'this' determinate figure in space. But this presupposes that there are independent criteria of *identification* for these determinate figures, though not necessarily criteria for making direct *identifying* references to them. After all, it is part of Newton's use of the distinction between the order of being and the order of knowing, to warn against confusing *what* is picked out with *how* it is picked out. This is surely what lies behind his claim, that in distinguishing absolute from relative space we must not "confound real quantities with their relations and sensible measures."[105] Thus the conditions under which an individual can be identifyingly referred to, are not the same as the criteria for its *being* an individual of that sort or kind. So while in the order of referring to individuals identifyingly, sensible determinations of the form of matter are prior to the figures in space, in the order of being these figures possess separate criteria of identification.

The trouble is that Newton fails to provide an account of these criteria which shows how they can be exemplified independently of sensible measures. Certainly, to distinguish identifying conditions from criteria of identification, does not show the latter to be independent of the former. However, Newton's explicit statement on the latter is this: "the parts in space [figures] derive their character from their positions, so that if any two could change their positions, they would change their character at the same time, and each would be converted numerically into the other." But this is absurd: the figures in space are "understood to be the same as they really are" because of the immutable ordering of their positions. And apart from that ordering they lack "any hint of individuality."[106]

Newton is doubtless attempting to specify some means of identifying the figured parts of space independently of sensible measures, and one which will avoid abvious circularity. Unfortunately, he confronts a notorious difficulty. There seems no way of analyzing the spatio-temporeal continuity of any bit of matter without quantifying over places and times. This means that the criteria of identity over time for a figured part in space depend on some bit of matter remaining numerically one and the same at different times. Equally, however, criteria of identity over time for a bit of matter presuppose that places in extended space remain numerically one and the same at different times. Newton cannot break this ontological circle by arguing that matter and space are mutually dependent. He unfortunately commits himself to the view that space (as well as time) is ontologically prior to, and really distinct in nature from, the form of matter that it instantiates.

IV. NEWTON'S CRITIQUE OF DESCARTES' THEORY OF MATERIAL SUBSTANCE

So much then, for Newton's conception of the nature of space and matter. But it remains to show how he consciously rejects Descartes' position in advancing his own. But first, what is Descartes' conception of matter and space? Since Newton's references in *De Gravitatione* are to Descartes' *Principia Philosophiae*, it will be well to outline in more detail the theory of material extension that is developed in that work. According to Descartes, the necessary *and* sufficient condition for something's being material in kind, is that it be extended in length, width and breadth.[107] To say, then, that something is extended is not merely to predicate a property which truly applies. Rather, it says what it is; namely, what being a material thing is essentially. Thus to say of x that it is extended is not at all like saying of x that it is red. In other words, a thing's extension is identical with its essential nature, and is thus something that is predicated essentially of that nature. There is thus only a distinction of reason between extension as the principal attribute of corporeal substance and substance itself. So matter is extended simply in virtue of the sort of thing that it is; not because the mind conceives that it signifies what the definition of being an extended thing entails. Furthermore, everything that can truly be said of an individual material thing presupposes extension, and is thus merely a mode of that thing's essential nature.[108] So while the mind can consider a particular shape in abstraction from the thing of which it is the shape, it cannot conceive that shape and deny that it is extended, anymore than it can deny that extension is the principal attribute of corporeal substance according to which its complete reality is comprehended.[109] Equally, motion geometrically considered is a relational mode of body in that it is completely characterized in terms of the distance the body translates with respect to an extended space determined by other bodies considered as being at rest.[110] As we might expect, Descartes denies that there is a real distinction between the essence of corporeal substance and that of space. *In re* both are tri-dimensionally extended and hence indistinguishable. Thus the essence of matter is extension and space is defined derivatively. Similarly, he holds that the "intrinsic place" of a thing is not really distinct from its extension. So when a thing is said to be in a place, there is no ground for the existence of a separate entity – place or derived generic space – that is distinct from the thing's existence.[111]

It is clear that Descartes' ontology is stark in the extreme and that it is based on the following principles. No real distinction exists between bodies and space; real places do not exist independently of the disposition of bodies; and lastly, all true properties of bodies are merely the proper and derived modifications of *res extensa*. Equally clearly, Newton rejects all three: there is a real distinction between matter and space; places are intrinsic to the structure of space and capable of being filled by matter; moreover, the properties of matter which distinguish it from space are not derived from extension, but are the product of God's active will. There can be little doubt that Newton consciously opposes these fundamental principles to those of Descartes.

However, it is necessary to consider the implication of this in more detail. In the following passage, Newton vigorously denies the Cartesian argument that when matter is stripped of all sensible and accidental qualities only extension remains as its essence:

> . . . see how fallacious and false this Cartesian argument is; for if indeed the accidents of bodies have been rejected, there remains not extension alone, as he imagined, but also powers by which they can stimulate perceptions in the mind and move other bodies. If we further reject these powers and all capability of moving so that there only remains a precise conception of uniform space, will Descartes fabricate any vortices, any world, from this extension? Surely not, unless he first involves God who alone is able to bring forth [*procreare*] bodies *de novo* in those spaces (by restoring those powers, or the corporeal nature, as I explained above). And so in what has gone before I was correct in assigning the corporeal nature to the powers already enumerated.[112]

We could not ask for a better summary of Newton's position and a clear indication of how he takes it to differ from Descartes' conception of material substance. As before, Newton identifies the form of matter – the corporeal nature – with powers directly created by Divine will. Thus the essential nature of matter involves not only extension as Descartes maintains, but active powers which intrinsically embody causal properties. For the "powers of bodies can be rejected without violating extension, but not without violating their corporeal nature."[113] Furthermore, Newton rightly observes that if in accordance with Descartes' ontology of nature only *res extensa* exists independently of perception, there is no possibility of generating a determinate world without the operation of Divine power. So even on Descartes' showing, it is Divine action that must directly inform space with active powers, and thus provide a metaphysical basis of matter. Moreover, in Newton's view, without these powers there would be no real basis in nature itself for change, motion and activity.

Though Newton is right to claim that by its nature Cartesian extension cannot embody either the power of changing in itself or of changing other things, he is wrong to suppose that Descartes thought otherwise. Indeed,

Descartes' view of the source of change is similar to Newton's. Like Newton he holds that causal action can only be predicated of conscious agents such as God and created minds. Thus the sole and true source of motion, change and activity is Divine omnipotence. In fact, Descartes' voluntarism is more extreme than Newton's, since he holds that God not only creates the world but recreates it in each successive slice of time.[114] All created things are utterly contingent, which for Descartes means that they do not embody in their defining nature the power of continuing in existence beyond the present moment. Since created things are contingent in this sense, and since they are brought into existence from nothing at a given moment of time, it follows that they must be recreated by God at each successive time.[115] It is clear that Newton lacks a complete grasp of the principles which support Descartes' thinking.

This is evident when he interprets Descartes' conception of *res extensa*. His main concern is to show that Descartes' view of matter leads to atheism while his own does not. Of his own position he says: "we cannot posit bodies of this kind without at the same time positing that God exists, and has created bodies in void space out of nothing, and that they are beings distinct from created minds, but able to unite with minds." On the other hand, "if we say with Descartes that extension is body, do we not manifestly lay down a path to Atheism, both because extension is not created but has existed from eternity, and because we have an absolute Idea of it without any relation to God, so that we are able to conceive extension while meanwhile imagining the non-existence of God?"[116] There can be no doubt that Newton is interpreting Descartes' conception of extension in the light of his own. According to Newton, extension to eternal, infinite and uncreated just because it is co-eternal with God and a direct consequence of his existence. It is this conception of extension which he identifies with Descartes' view; namely, that the essence of matter is to be extended. Little wonder, then, that Newton is concerned to overthrow the Cartesian theory of matter. For as he conceives Descartes' theory, material reality is not only eternal and uncreated, but co-eternal with the existence of God. So interpreted, it follows immediately that Descartes' account of the present material world makes it independent of God's power and providence. However, Descartes' true position conceives material extension as a dependent substance continually recreated by Divine power. Thus, instead of Newton's conception of extension as the eternal and uncreated space of Divine presence, in Descartes' view material extension (with space defined derivatively) is a created being which continuously owes its nature and existence to the will of God.

But this is not the only way in which Newton misunderstands Descartes' theory of matter. He observes acidly that we must lay "aside that puerile prejudice derived from infancy according to which extension is inherent in bodies like an accident in a subject without which it cannot truly exist."[117] It is clear that Newton is reiterating his view that extension is neither a substance, in the sense of a substratum, nor an accident. But he is also implying that Descartes conceives the attribute of extension as dependent on material substance in the way which accidents are said to be dependent particulars according to scholastic thought. But this is not Descartes' view. The intelligibility of extension is grasped immediately by the mind; and no separate inference is needed to the existence of an underlying substantial support in which it inheres. Rather, when extension is comprehended, the essence of material substance is directly known in itself, since there is only a distinction of reason between any substance and its principal attribute.[118] So when Newton castigates those who substitute the substantial reality "of attributes which are real and intelligible things in and of themselves" for the hidden reality of an unknowable substratum, his aim misses Descartes. In Descartes' view, extension is intrinsically intelligible in and of itself; and in grasping it the mind knows *ipso facto* the essential reality of material substance.

V. CONCLUSION

There are a number of lessons to be learned from Newton's *De Gravitatione*. Of course, it is the work of a young man; but it advances an original and interesting metaphysical theory regarding the nature of space and the creation of matter. Here we witness the birth of the theory of absolute space whose properties and relations Newton holds to exist independently of non-spatial things. And while he misunderstood the basis of Descartes' doctrine of material extension, it is clear that Newton's response to the latter's views contributed to the articulation of his own conception of extension. In fact, the response to Descartes, within an Epicurean framework, probably has more relevance to the ultimate form which Newton's conception of extension takes, than any particular influence based on his knowledge of Henry More's conception of space.[119] After all, according to More's mature conception, space is an incorporeal substance which grouds the same descriptions that apply to Divine presence.[120] But it is just this conception of the nature of extended space in relation to Divine presence that Newton wishes to reject. Space is not an attribute of Divine being, as such, but a consequence of God's omnitemporal existence.

Equally, it is clear that Newton's conception of matter as a set of powers with which space is informed by God, is a result of his reading atheistical implications into Descartes' theory of material extension. Moreover, Newton's view of matter as an essentially active substance (which he takes to be a basic feature of corporeal nature) was to remain, transformed to be sure, in his later thought. In *Principia* (1687), the conception of matter as possessing a *vis insita* which, though lacking the capacity to initiate new motion, preserves a body's state, and also passively resists its being put out of that state, is a descendent of Newton's discussion in *De Gravitatione*.

Lastly, it is clear that the origins of Newton's theory of matter and of space cannot be appreciated without a grasp of the intellectual traditions which he accepted and rejected. As his early notebook and *De Gravitatione* testify, Newton concerned himself with philosophical issues. If we are to understand him fully, we must do so as well.

University of Pittsburgh

NOTES

* The present study was first written over four years ago. While I still agree with its overall interpretation, there are claims I should now modify. Specifically, I should now place less stress on Newton's incorporating indivisible minima into the structure of space in *De Gravitatione*; I now believe that he had given up that program before writing that treatise. I should also treat the problem of the unity of space differently, though not essentially differently; and I should now argue more emphatically for Newton's insisting upon a Platonic distinction between natures and natured things in his discussion of extension and extended things. These issues are dealt with in a current study, 'Space, Geometrical Objects and Infinity: Newton and Descartes on Extension'.

[1] This is transcribed and translated by A. Rupert Hall and Marie Boas Hall, in *The Unpublished Scientific Papers of Isaac Newton*, (Cambridge University Press, New York 1962), 90–156. (Hereafter, *De Gravitatione*. I shall indicate when I have changed the Halls' translation). Newton did not give his treatise the title *De gravitatione et aequipondio fluidorum*. This phrase in fact comprises the first five words of the first sentence of the treatise. Since first published by the Halls, however, it has served as a title.

[2] Isaac Newton's *Philosophiae naturalis principia mathematica* (fasimile of third edition (1726) ed. A. Koyré and I. B. Cohen), Vol. I (Cambridge, 1972). Scholium on Space and Time, 46–53. All citations are from this edition. (Hereafter cited Koyré and Cohen).

[3] The first part of *De Gravitatione* is a systematic critique of Descartes' *Principia philosophiae*. Descartes' influence on Newton is not merely negative. Rather, Newton's own position has a character it otherwise would not have had, if he had not examined the views of Descartes.

[4] *Oeuvres de Descartes*, ed. Charles Adam and Paul Tannery, Vol. 8, Pars Secunda, Article 4. (hereafter, AT.)

5 *De Gravitatione*, p. 140 and p. 144.

6 *Ibid*., 144. I have slightly altered the translation. See also p. 143 for another expression of this view. See Martha Brandt Bolton, 'Substances, substrata, and Names of substances in Locke's *Essay*', *Phil. Review* **65** (1976), 488–513.

7 AT., Vol. 8, pars secunda, Article 10, 45.

8 If we say with Descartes that matter *is* extension, and with Newton that extension is uncreated, matter is then independent of Divine creation. Newton states that extension is uncreated at p. 145 *De Gravitatione.*

9 That is, uncreated extension is knowable in and of itself because its uncreated nature does not depend on anything other than itself.

10 *De Gravitatione*, 145.

11 *Ibid*., 137, translation mine. Latin on page 104.

12 For a full discussion of this doctrine see J. E. McGuire, 'Existence, actuality and necessity: Newton on space and time', *Annals of Science* **35** (1978), 463–508. Hereafter 'Existence, actuality and necessity'.

13 In other words, for Newton Divine presence is explained in a way that is akin to the Medieval doctrine of how something is definitively in place. A thing is definitively in place if it is whole in the whole place and whole in each part of place. Given these assumptions, it would be absurd to say that God has continuously to bring it about that he is definitively in place.

14 *De Gravitatione*, 139.

15 Cambridge University Library, Ms. Add. 3996, folio 87^{r} (p. 131. The brackets are Newton's (hereafter *Questiones*).

16 *De Gravitatione*, 132–133 (includes the quoted passage). Following the Aristotelian conception of definition, Newton says that he will establish *that* space *is*, and then *what* it is. Not only is Newton talking his place in the Aristotelian tradition on definition, but he is consciously rejecting Descartes' theory of definition as found in the *Regulae*. Newton's argues for his view of the actual infinity of space, by means of geometrical examples, in *De Gravitatione*, 133–136. In so doing, he carefully examines Descartes' claim that generic extension is indefinite in nature. The linking of prefection, infinity and actuality is a Christian notion of the fullness of being, which has roots in the thought of Anaximander and the Ancient atomists.

17 *De Gravitatione*, 132–133. I have altered the translation slightly.

18 *Ibid*., 122.

19 *Questiones*, folio $88^{r\text{-}v}$–$89^{r\text{-}v}$, $119^{r\text{-}v}$–120^{r}. Newton's discussion is well grounded in Ancient, Medieval and contemporary Seventeenth century sources. The proximate sources for Newton's Epicurean views on indivisible magnitude is Walter Charleton's *Physiologia Epicuro-Gassendo-Charltoniana*, (London, 1654). It is highly probable, however, that Newton knew Diogenes Laettius directly. On things Epicurean, I profited from David J. Furley's *Two Studies in the Greek Atomists*, Study I, chapters 5, 6, 7 and 8 (Princeton University Press, 1967).

20 *Questiones*, folio 89^{r}.

21 The problem is this. If something is actually infinite, it cannot be said to have internal differences and distinctions. Nor can it be limited by any finite mark. But in Newton's view, infinite extension possesses minima which, though indivisible in nature, nevertheless qualify its overall infinitude. This difficulty is discussed in detail below.

22 *Questiones*, folio 89^{r}, $119^{r\text{-}v}$, 120^{r}.

[23] *Physics*, (Leob edition), Bk III, chp. VI, (207a10–15, 207b1–25).
[24] Suffice it to say that they deal with Zenonian and Aristotelian objections to the divisibility of magnitude into indivisible parts.
[25] Koyré and Cohen, 46. One of the intractable difficulties in Newton's early position on space is whether it is compatible with the principles of geometry (or conversely for that matter). This raises a large number of questions which cannot be dealt with here. But problems of incommensurability, irrationality and the infinite divisibility of magnitude, all of which were raised by the geometers against the Epicureans, seem to plague Newton's claim that space possesses a structure of inherent and indivisible parts.
[26] See Edward Grant, 'Place and space in medieval physical thought', in *Motion and Time, Space and Matter*, ed. by Peter K. Machamer and Robert G. Turnbull (Ohio State Press, 1976), 137–167. The *locus classicus* for the view which connects corporeality and extensionality is Aristotle's *Physics*, Bk IV. For Aristotle, extension is the extension of corporeal individuals and without them there is no space.
[27] AT., Vol. 8, Pars Secunda, Articles 18, 50.
[28] *Ibid*., Articles 16, 17, 18, 49–50.
[29] *Ibid*., Articles 10, 11, 12, 13, 45–47.
[30] *De Gravitatione*, 132.
[31] *Ibid*., 132. I have altered the Halls' translation; words and phrases in brackets have been added.
[32] See AT., Vol. 19, *Regulae* XIV, 57–59 and *Principia*, Vol. 8, Pars Secunda, Article 9, 45.
[33] *De Gravitatione*, 132. Translation mine. Latin on page 99.
[34] *Op. cit.*, 'Existence, actuality and necessity', 474–484.
[35] *Ibid*., 465–469.
[36] *De Gravitatione*, 103. The translation is mine.
[37] AT., Vol. 8, Pars Secunda, Articles 11, 12, 13 and especially 14, 45–47.
[38] A. Koyré, and I. B. Cohen, 'Newton and the Leibniz-Clarke Correspondence', *Archives inter. d'hist. sci*., nos. 58–59 (1962), 99. These and related passages are fully discussed in 'Existence, actuality and necessity'.
[39] In Newton's view God is not only omnitemporal, but he exists at all places in space.
[40] *De Gravitatione*, 138.
[41] *Ibid*., 135–138.
[42] *Ibid*., 132. Translation mine. Latin page 99.
[43] *Ibid*., 99. Translation mine.
[44] Henry More, *Enchiridion Metaphysicum*, (London, 1671), Part I, Cap. I, Section 7, and Cap. 6, Sections 1, 2 and 3. More argues for the same position in his *Divine Dialogues*, (London, 1668), Dialogue, I, 96. Newton could have read this.
[45] *Ibid*., Part I, Cap. 38, Section 10, 11 and 13.
[46] *Enchiridion*, Part I, Cap. 2, Section 2.
[47] Francesco Patrizi, *Nova de Universis Philosophia* (Ferrara, 1591). See section entitled *De Spacio Physico.*
[48] Blaise Pascal, 'Concerning the vacuum', Letter from Pascal to M. Le Pailleur, October, 1647, *Great Books of the Western World*, ed. R. M. Hutchins, (University of Chicago Press, 1952), p. 370. I have not been able to consult the original.
[49] *De Gravitatione*, 137.
[50] *Ibid*., 134.

[51] Descartes allows that this distinction applies only to the infinity of Divine nature, but it does not partain to extension. See A. Kenny, *Descartes' Philosophical Letters*, (Oxford, 1970), 245.

[52] *De Gravitatione*, 136. I have slightly altered the translation.

[53] *Ibid.*, 135. For a full discussion of Newton's view of how perfections and infinity relate see 'Existence, actuality and necessity'. The view that completed perfections pertain to that which is unlimited and infinite clashes with the Greek view that perfection equates with form as something which is determinate, limited and thus definable. This tension remains in theological and philosophical thought.

[54] *Ibid.*, 134–135.

[55] AT., Vol. X, *Regulae*, see regulae 12 and 14, 410–430, 438–452.

[56] Benedict Spinoza, *Opera*, Ed. J. Van Vloten and J. P. N. Land, (Hague, 1924) third edition, Vols. I–II, *Ethics*, pars prima, Prop. XV, Scholium, 48–51.

[57] *Ibid.*, pars prima, Definition III, 37.

[58] *Ibid.*, pars prima, props. XIX, XX, and XXI, 54–56. Spinoza says explicitly that there is only a distinction of reason among the attributes of Divine being. Presumably he means that the human mind is compelled to make these distinctions, and that it is the reality of Divine being itself which forces them to be made. What he does not say explicitly is whether it is the activity of the intellect alone that distinguishes Divine attributes, or whether there exists *in re* some prior ground for their being distinguished. For a fine discussion of the theory of distinctions see Marilyn McCord Adams, 'Ockham on Identity and Distinction' *Fransiscan Studies* **26**, Annual XIV, (1976), 5–74.

[59] *Ibid.*, pars prima, Scholium to Prop. XV.

[60] *Ibid.*, pars prima, Def. VI, 35.

[61] *Ibid.*, pars prima, Prop. XI, 43–46.

[62] *Ibid.*, also see explanation to Def. VI, 37.

[63] *Ibid.*, Prop. X, 43.

[64] *Ibid.*, Prop. XV, Scholium 48–51.

[65] *Ibid.* That there is a systematic connection between the infinity and indivisibility of any form of being is argued by Proclus. See *The Elements of Theology*, trns. E. R. Dodds, (Oxford, 1963), prop. 86. However, like Spinoza he holds that it is only principles or self-subsisting entities that can provide a ground for connecting indivisibility with an infinite nature.

[66] *Ibid.*, see also the letter to Meyer, 20 April, 1663, *Opera*, Vols. III–IV, 38–43.

[67] For a discussion of levels of knowledge in Spinoza see James Lennox, 'The causality of finite modes in Spinoza's ethics', *The Canadian Journal of Philosophy* 6 (1976), 479–500. Spinoza's distinction between rational and intuitive knowledge, in the context of levels of knowledge, is relevant, but it does not change the character of the position here developed. On this position see Spencer Carr, 'Spinoza's distinction between rational and intuitive knowledge', *Phil. Review* (April, 1978), 241–252.

[68] See discussion above.

[69] Spinoza, *Opera*, Vols. I–II, Prop. XV, 50.

[70] For a discussion of this point see 'Existence, actuality, and necessity'.

[71] *Ibid.*, and also see note 13 above.

[72] See note 24 above.

[73] *Opera*, Vols. I–II, *Tractatus de Intellectus Emendatione*.

[74] See note 16 above.

[75] *De Gravitatione*, 132.
[76] *Ibid.*, 95–96.
[77] *Ibid.*, 132.
[78] This is a position (developed from Aristotelianism) which says that while we can grasp the idea of a finite magnitude to which proportionate units are added endlessly, there is no idea of an actually infinite magnitude. For Aristotle's more subtle discussion of infinite magnitude see Book III of *Physics*, 200b30–208a. For a seventeenth century discussion of this view of magnitude and quantity see John Locke, *An Essay Concerning Human Understanding*, (New York, Everyman's Library, 1974), Vol. I, Book II, Chapter XVII, 171–184. It seems that Newton alternates between conceiving space as either a magnitude which is a negation of finite extension, or as an infinitely extended expanse without boundaries, beyond which nothing can exist.
[79] In the Scholium on space and time of the *Principia*, Newton speaks of space as a magnitude which has parts; however, he does not say explicitly, as he does earlier in *Questiones* and *De Gravitatione*, that space actually possesses parts that are intrinsic to its structure. Nevertheless, he certainly supposes that space has a natural metric which is based on the internal structure of space itself. As against this, however, he argues in a document written in 1692 that "space itself has no parts which can be separated from one another, or be moved among themselves, or be distinguished from one another by any inherent marks. Space is not compounded of aggregated parts since there is no least in it, no small or great or greatest, nor are there more parts in the totality of space than there are in any place which the very least body occupies. In each of its points it is like itself and uniform nor does it truly have parts other than mathematical points, that is everywhere in number and nothing in magnitude." It seems here that Newton is denying that space is a magnitude in any sense whatsoever. Moreover, the passage supports Newton's standard claim that space is an indivisible entity. This is, however, the only place where he explicitly grounds spatial structure on the indivisibility of mathematical points. In any event, the passage contributes to the fact that Newton's thought concerning the structure of space changes (at least in 1692) from the time of the *Questiones* to the post-*Principia* period. These changes have yet to be documented and analyzed. For a full transcription of the 1692 document see J. E. McGuire, 'Newton on place, time, and God: an unpublished source', *B.J.H.S.* **11** (1978), 114–129.
[80] For a discussion of this see J. E. McGuire, 'The origin of Newton's doctrine of essential qualities', *Centaurus* **12** (1968), 223–260.
[81] Koyré and Cohen, *Principia*, Vol. I, 49.
[82] *De Gravitatione*, 133.
[83] *Ibid.*
[84] *Ibid.*, 138.
[85] *Ibid.*, 139.
[86] *Ibid.*, 139.
[87] *Ibid.*, 130. I have altered the Halls' translation.
[88] *Ibid.*, 140.
[89] *Ibid.*, 142. Newton rejects the need for intermediary entities such as the soul of the world.
[90] *Ibid.*, 141.
[91] *Ibid.*, 141.

92 For a discussion of Newton's voluntarism see J. E. McGuire, 'Force, active principles and Newton's invisible realm', *Ambix* **15** (1968), 154–200.

93 *Ibid.*

94 *De Gravitatione*, 140.

95 *Ibid.*, 132.

96 *Ibid.*, 141–142.

97 *Ibid.*, 142.

98 *Ibid.*, 140. Translation mine. Latin on pages 106–107.

99 *Ibid.*, 140–141.

100 Aristotle's views on prime matter are still a controversial issue. See Aristotle's *Physics* I, II, translated by W. Charlton, Appendix, 'Did Aristotle believe in prime matter', 129–145. For a rejection of Charlton's claim that Aristotle did not believe in a qualityless substratum, see H. M. Robinson, 'Prime matter in Aristotle', *Phronesis* **19** (1974), 168–188.

101 *De Gravitatione*., 148.

102 *Ibid.*, 140.

103 *Ibid.*, 138–139.

104 *Ibid.*, 146.

105 Koyré and Cohen, *Principia*, Vol. I, 52.

106 *De Gravitatione*, 136.

107 AT., Vol. 8, pars secunda, Article 4, 42.

108 AT., Vol. 8, pars secunda, Article 11, 46.

109 AT., Vol. 3, 472–480. Descartes to Gibieuf, 19 January, 1642.

110 AT., Vol. 8, *Principia*, pars secunda, Article 25, 53.

111 *Ibid.*, Article 10, 45.

112 *De Gravitatione*, 147–148. I have altered the translation.

113 *Ibid.*, 146.

114 AT., Vol. 8, *Principia*, pars prima, Article 20, 12–13 and Vol. 7, Mediation III, 48–49.

115 *Ibid.*

116 *De Gravitatione*, 142–143.

117 *Ibid.*, 138. Translation mine. Latin on page 105.

118 AT., Vol. 8, *Principia*, pars prima, Articles 52 and 53, 24–25. Descartes often speaks of substance as the substratum of the attributes that specify it (AT., Vol. 7, 285–286, AT., Vol. 5, 156, AT., Vol. 9, 47). He never doubts, however, that the *essential* nature of a substance is grasped through its principal attribute. Of the underlying substratum itself we can have but a confused perception.

119 See 'Existence, actuality and necessity'.

120 *Divine Dialogues, Dialogue* I, 10.

R. W. HOME

NEWTON ON ELECTRICITY AND THE AETHER*

It is well known that Newton became convinced towards the end of his life that electricity played a vital role in the operations of nature. In the famous final paragraph of the *Scholium Generale* that he added to the second edition of the *Principia*, published in 1713, he wrote of "a certain most subtle spirit which pervades and lies hid in all gross bodies." It was this active spirit that gave rise, he supposed, to the electrical attractions and repulsions that manifested themselves at sensible distances from most bodies after they had been rubbed, as well as to the cohesion of particles when contiguous. In addition, he surmised, it was the agency responsible for the emission, reflection, refraction, inflection and heating effects of light; and by its vibrations in "the solid filaments of the nerves," it carried sensations to the brain, and commands of the will from the brain to the muscles in order to bring about bodily motion.[1]

In Andrew Motte's 1729 English translation of the *Principia*, the electrical character of this strange substance was emphasized by the substitution of the phrase "this electric and elastic spirit" for Newton's unadorned "this spirit" (*hic spiritus*) in the final sentence. Yet Motte was by no means misrepresenting Newton's views here: on the contrary, the phrase *spiritus electricus* appears in one of the draft versions of the *Scholium Generale*,[2] and Newton later wrote in the margin of his own annotated copy of the second edition of the *Principia*, after the word *Spiritus*, the words *electrici & elastici.*[3] Moreover, the electrical character of the subtle spirit was already clear in the original published version of the Scholium, from the role Newton attributed to it of bringing about the motions traditionally subsumed under the rubric "electricity".[4] Being electrical, it is clearly not to be equated with the spirit of God, as Koyré once suggested it might be. Instead, it must be located unambiguously among the ranks of second causes.[5]

Soon after the publication of the *Scholium Generale*, Newton's views concerning the subtle spirit underwent some changes. These are reflected in the eight additional Queries, numbered 17–24, that he included in the new edition of his *Opticks*, published in late 1717.[6] In certain respects, the "Æthereal Medium" that he described in these Queries was like the "subtle spirit" of the earlier *Scholium Generale*. It shared with the subtle spirit such properties as rarity, subtlety, activity and elasticity. Like the subtle spirit, it

Z. Bechler (ed.), Contemporary Newtonian Research, 191–213.

was held responsible for the reflection, refraction, inflection and heating effects of light, and for the functioning of the nervous system in transmitting sensations from the sensory organs to the brain, and commands of the will from the brain to the muscles. No longer, however, was it said to be chiefly confined, as the "subtle spirit" was, to be pores of "gross bodies." On the contrary, Newton suggested that this new medium was by its elastic force "expanded through all the Heavens" and was densest, he thought, in spaces altogether devoid of ordinary matter. It was therefore a true aether, something the earlier "subtle spirit", being more localised, was not. The difference is important, and tends to be reflected in Newton's terminology.[7]

Henry Guerlac has argued that Newton's reintroduction of a genuine, universally disseminated aether at this point was inspired by an experiment performed before the Royal Society of London in 1716 by Jean Théophile Desaguliers, at Newton's instigation.[8] This experiment, on the comparative rates of cooling of thermometers in air and in vacuo, was described by Newton in one of the new Queries (Q. 18) added to the *Opticks* a year later; and Guerlac has drawn attention to a letter written by Desaguliers to Sir Hans Sloane many years afterwards, in which it is said in so many words that the experiment was carried out at Newton's behest.

In the course of the "aether" Queries, Newton applied his new conception to the explanation of optical phenomena in particular. Most of those which he had previously accounted for in terms of the "subtle spirit" were now said to depend upon the aether, and so, too, were the mysterious "fits" of easy reflection and easy transmission of light that he had discovered long before, but had always tended previously to leave unexplained. These, he now suggested, had their origin in vibrations excited in the aether by rays of light falling on the surfaces of pellucid bodies.[9]

It is notorious that elsewhere in the new Queries, in Query 21, Newton offered an explanation of gravity in terms of density gradients within his aether.[10] Some scholars have taken a similar explanation, in terms of the "subtle spirit," to be implicit in the *Scholium Generale,*[11] but this is to count for nought Newton's unambiguous statement in the immediately preceding paragraph of the Scholium that he feigned no hypotheses concerning the cause of gravity.[12] Furthermore, earlier drafts of the Scholium reveal that Newton in fact maintained a clear distinction between gravity, on the one hand, and the various effects he listed among those which would find their explanation in terms of the "subtle spirit."[13] With the apparent exception of electricity (and we shall see that in Newton's mind even this was in fact no exception), the latter all depended upon short-range forces only; apart from electricity, they occurred in the immediate vicinity of bodies, precisely where Newton said the "subtle spirit" was to be found. Newton rightly regarded gravity as being in a quite different category. If it was to be explained mechanically, a mechanical agent was required that extended, as the "subtle spirit" did not, over the entire space between widely separated material bodies. In other words, only when he replaced the "subtle spirit" by a universally disseminated medium did such an explanation become a possibility so far as Newton was concerned. And even after this, so Desaguliers tells us, Newton remained hesitant about identifying the "Medium acting upon light" with whatever it was that caused bodies to gravitate towards one another.[14]

The replacement of the "subtle spirit" of the *Scholium Generale* by the universal aether of the late Queries in the *Opticks* was attended by other, less striking changes in the explanatory role attributed to it by Newton. In particular, several of the phenomena that were said in 1713 to be explicable in terms of the "subtle spirit", namely cohesion, electrical attraction and repulsion, and the emission of light from luminous bodies, were not mentioned in 1717 among those brought about by aetherial action.

All these omissions were surely deliberate. Cohesion and the emission of light were, in Newton's general scheme of things, precisely the phenomena that depended upon the most intense and most short-range forces of all, one attractive, the other repulsive. Among the universal or quasi-universal forces of matter, these differed most from gravity in their mode of operation, and were therefore the least likely to be brought about by the same causal principle.

The omission of electricity was certainly intentional, as we learn from a passage in one of the newly inserted Queries, Query 22. Here, Newton confronts the question of how the motions of the heavenly bodies can continue undisturbed in an all-pervading medium such as the aether he is here invoking. His solution is to suppose that his aether is both extremely rare and extremely elastic; and to give this suggestion at least a modicum of plausibility, he cites instances already "known" in which matter retains its activity when reduced to a very low density. What is significant for us here is that one of his examples is the subtle matter which, he assumes, gives rise to electrical motions. "If any one would ask how a Medium can be so rare," he says, "let him ... tell me, how an electrick Body can by Friction emit an Exhalation so rare and subtile, and yet so potent, as by its Emission to cause no sensible Diminution of the weight of the electrick Body, and to be expanded through a Sphere, whose Diameter is above two Feet, and yet to be able to agitate and carry up Leaf Copper, or Leaf Gold, at the distance of above a Foot from the electrick Body?"[15] There is no suggestion here that the aether and this electrical "exhalation" are one and the same thing. On the contrary, the electrical exhalation is taken for granted as something different, whose nature is well known and understood, and whose properties provide a useful analogy to buttress Newton's account of the aether. Unlike the "subtle spirit" of the *Scholium Generale*, the aether of the *Opticks* Queries is not an electrical spirit. By 1717, Newton's flirtation with the notion that electricity was a universal causative agency in the micro-realm was apparently at an end.

Or was it? It is plain from the passage just quoted that Newton's belief that electrical motions were brought about by *a* subtle spirit emitted by electric bodies was in no whit diminished. Furthermore, it was well known in his day that many if not all bodies could be rendered electrical by friction. All these bodies, at least, must therefore contain some kind of subtle and active matter that could it appropriate circumstances be extruded from them. In addition, the question of what caused cohesion and the emission of light remained unanswered. It seems likely that Newton continued to hold that these effects were brought about by an electric spirit that pervaded the pores of all gross bodies. Even if he did not, however, he certainly continued to believe in the existence of the electric spirit itself, as will be made abundantly clear in what follows. Hence, the universally disseminated aether of the 1717 *Opticks* Queries ought not to be seen as a transmogrification of the more localised electric spirit of 1713. Rather, it was a *second* active medium introduced into the discussion at this point, to which certain of the functions previously ascribed to the electric spirit had been transferred, and to which various others had been attributed in addition.[16]

This interpretation, if correct, prompts an obvious question. If, in 1717, Newton really saw the electric spirit and the aether as two different kinds of active medium, why did he not make this clearer to readers of the new edition of *Opticks*? Certain unpublished drafts of proposed additions to the Queries suggest that he initially intended to do so. In the end, however, his habitual caution won out, and the relevant passages were suppressed.

At one point in the preparation of the new edition, Newton brought together on a single sheet a series of insertions he intended to introduce into the original set of Queries published in 1704. Included amongst these was the well known passage, destined for Query 8, describing some of Francis Hauksbee's striking experiments with electrified glass globes. Almost all the proposed insertions were in due course included without further amendment in the new printing. The following, however, which Newton at first intended to insert as a new Query at "Pag. 134 lin 1" (that is, immediately after Query 8), was omitted:

Do not electric bodies by friction emit a subtile exhalation or spirit by which they perform their attractions? And is not this spirit of a very active nature & capable of emitting light by its agitations? And may not all bodies abound with such a spirit & shine by the agitations of this spirit within them when sufficiently heated? ffor if a long cylindrical piece of Ambar be rubbed nimbly it will shine in the dark & if when it is well rubbed the finger of a man be held neare it so as almost to touch it, the electric spirit will rush out of the Ambar with a soft crackling noise like that of green leaves of trees thrown into a fire, & in rushing out it will also push against the finger so as to be felt like the ends of hairs of a fine brush touching the finger. And the like happens in glass. If a long hollow tube of flint glass about an inch be rubbed nimbly with a paper held in the hand till the glass grows warm, it will in rubbing emit light & the face or any other tender part of the skin being held neare that part of the glass where it has been most rubbed, the electric spirit which is excited by the friction will rush out of the glass with a cracking noise & push against the skin so as to be felt, & in pushing emit light so as to make the skin shine like rotten wood or a glow worm. And if the glass was held neare pieces of leaf brass scattered upon a table the electric spirit wch issued out of the glass would stir them at the distance of 6, 8 or 10 inches or a foot, & put them into various brisk motions, making them sometimes leap towards the glass & stick to it, sometimes leap from it with great force, sometimes move towards it & from it several times with reciprocal motion, sometimes move in lines parall (*sic*!) to the tube, sometimes remain suspended in the air, & sometimes move in various curve lines. Which motions shew that this spirit is agitated in various manners like a wind. And if a broad plate of glass be placed between the tube & the pieces of brass, yet the Tube will attract them them (*sic*!), tho not so strongly as when the plate of glass is taken away. Which shews that the electrick spirit is so subtile as readily to pass through glass tho not so readily as through the Air. And whilst it pervades dense bodies so easily, why may it not be latent in them all in some measure or other, tho those only emitt it by friction in which it abounds most copiously? And since it easily emits light by agitation, why may it not emit light in all dense bodies heated red hot & thereby cause them to shine?[17]

No clearer statement could be desired of the rationale underlying Newton's linking of the emission of light with the activity of the electric spirit, nor of the grounds on which he concluded that this spirit was present in greater or lesser amounts in all gross matter. The experiments upon which Newton bases his arguments were due, like those with the glass globe, to Hauksbee, and date from the period 1706–1709. So far as its contents are concerned, therefore, the new Query could have been drafted at any time between about 1710 and late 1717. Yet the fact that the other passages on this page had reached their final form by the time the page was written out suggests that the whole was of relatively late composition, and perhaps even post-dated Newton's revived interest in a universal aether as an explanatory mechanism. More conclusive is the fact that Newton subsequently re-numbered his draft Query as "Qu. 18B." Since a Query of this kind would have been quite out of place between Queries 18 and 19 of the 1706 Latin *Optice* (that is, between Queries 26 and 27 in the 1717 English version), it must be that Newton intended it to go between the new Queries 18 and 19 of 1717, that is, between two of the early "aether" Queries in that edition. In other words, the re-numbering must have been done *after* the aether Queries themselves were drafted.

The extra Query would, in fact, have fitted quite logically at this point. It is precisely in Query 18 that Newton described Desaguliers' thermometer experiment and argued on that basis for the existence of a subtle medium, or aether, "expanded through all the Heavens."[18] There is not the slightest hint in the proposed Query 18B that the electric spirit discussed there was connected in any way with this universally disseminated agent. Rather, the new Query would have served immediately to introduce Newton's second subtle medium, an electric spirit that was also responsible for the emission of light from bodies under a variety of circumstances. In other words, it would have made plain his belief that there were in fact two different subtle media at work in the world.

For some reason, however, Newton was still not satisfied, and on the second of the pages into which the sheet was divided, he tried again, this time numbering it "Qu 24" and placing it after the last of the "aether" Queries and before the first of the re-numbered Queries from the Latin *Optice* (Query 22 of the eventual printed version was not included at this stage, so that the final "aether" Query was numbered 23, not 24):

Qu 24. Do not electric bodies by friction emit a subtile exhalation or spirit by wch they perform their attractions? And is not this Spirit – – – & thereby cause them to shine. Qu: 25 Are there not other original properties of light . . .[19]

Another attempt, which Newton numbered "Qu. 23", is recorded after this on the same page. It concentrates upon the detailed mechanism underlying electrical attraction and repulsion, and was presumably intended by Newton to preface his discussion in the proposed "Qu 24" of the involvement of electricity in the production of light. It includes an explicit suggestion that the electric spirit may be but one of several "subtle invisible Mediums which may have considerable effects in the Phaenomena of nature":

Qu. 23. Is not electrical attraction and repuls; performed by an exhalation wch is raised out of the electrick body by friction & expanded to great distances & variously agitated like a turbulent wind, & wch carrys light bodies along with it & agitates them in various manners according to its own motions, making them go sometimes towards the electric body, sometimes from it & sometimes move with various other motions? And when this spirit looses its turbulent motions & begins to be recondensed & by condensation to return into the electric body doth it not carry light bodies along with it towards the Electrick body & cause them to stick to it without further motion till they drop off? And is not this exhalation much more subtile than common Air or Vapour? For electric bodies attract straws & such light substances through a plate of glass interposed, tho not so vigorously. And may there not be other Exhalations & subtile invisible Mediums which may have considerable effects in the Phaenomena of nature?

And the same things have been found by rubbing a long & large cylinder of glass or Ambar with a paper held in ones hand, & continuing the friction till the glass grew warm.[20]

That electricity continued to be linked, so far as Newton was concerned, with the general phenomenon of cohesion is suggested by his choosing to append at this point a somewhat modified opening sentence for Query 31:

Qu. 31. May not the mutual attraction of the small parts of bodies be sufficiently strong to make them cohere & compose very hard bodies.

In the end, he did not use this, but he did add to the final sentence of the first paragraph of the same Query a remark that has since puzzled many an historian: "and perhaps electrical Attraction may reach to such small distances, even without being excited by Friction."[21] At this stage, he seems finally to have abandoned the idea of including an extended Query on the electrical spirit in the new edition of his book. In the end, all that remained was the brief allusion to an electric "exhalation" quoted earlier.

The significance of the document just described has been obscured by the presence in the same box of Newton's papers of a second set of draft Queries, also dealing with electrical matters, of which the first is again numbered 24. Quite understandably, there has been a tendency to assume that the two sets were connected at the drafting stage.[22] I believe, however, that they were not, and that the second set was in fact drafted somewhat earlier, probably at about the same time as the *Scholium Generale*. It reads as follows:

Quaest. 24. May not the forces by wch the small particles of bodies cohere & act upon one another at small distances for producing the above mentioned phænomena of nature, be electric? ffor altho Electric bodies do not act at a sensible distance unless their virtue be excited by friction, yet that virtue may not be generated by friction but only expanded, for the particles of all bodies may abound with an electric spirit wch reaches not to any sensible distance from the particles unless agitated by friction or by some other cause & rarefied by the agitation. And the friction may rarefy the spirit not of all the particles in the electric body but of those only wch are on the outside of it: so that the action of the particles of the body upon one another for cohering & producing the above-mentioned phænomena may be vastly greater then that of the whole electric body to attract at a sensible distance by friction. And if there be such an universal electric spirit in bodies, certainly it must very much influence the motions & actions of the particles of the bodies amongst one another, so that without considering it, philosophers will never be able to give an account of the Phænomena arising from those motions & actions. And so far as these phænomena may be performed by the spirit wch causes electric attraction it is unphilosophical to look for any other cause.

Quaest 25 Do not all bodies therefore abound wth a very subtile active potent elastic spirit by wch light is emitted refracted & reflected, electric attractions & fugations are performed, & the small particles of bodies cohære when contiguous, agitate one another at small distances & regulate almost all their motions amongst them selves, ffor electric – – – uniting the thinking soul & unthinking body. This spirit may be also of great use in vegetation, wherein these things are to be considered, generation, nutrition & præparation of nourishment. . . .

The Query then goes on to consider these aspects of living matter at some length.

In the passage quoted, Newton's dashes again indicate that he intended to take over *in toto* at this point a passage he had already written out elsewhere. The missing passage has been preserved as an un-numbered draft Query on the second half ot the sheet, and turns out to be quite extensive:

Quaest: Do not all bodies abound with a very subtil active vibrating spirit by wch light is emitted reflected & refracted, electric & magnetic attractions & fugations are performed, the small particles of bodies cohære when continguous, agitate one another at small distances & regulate almost all their motions amongst themselves as the great bodies of the Universe regulate theirs by the power of gravity? ffor electric bodies could not act at a distance without a spirit reaching to that distance. And by several experiments shewn by Mr Hawksby before ye R. Society it appears that a cylindrical rod of glass or hard wax strongly rubbed emitts an electric spirit or vapour wch pushes against the hand or face so as to be felt, & upon application of the finger to ye electric body crackles & flashes, & that the electric spirit reaches to ye distances of half a foot or a foot from the glass or above & passes readily through the solid body of a plate or vessel of glass, the electric body attracting things beyond the glass; & that if a globe of glass be nimbly turned round upon an axis & in turning rub upon a man's hand to excite its electric virtue, the the (*sic*) hand if the glass be empty of air shines through the glass wth a purple light, if some air be let into the glass, the whole cavity of ye glass appears illuminated wth flashes of a whiter light; if the Air be let in freely the glass emitts an electric vapour or spirit wch may be felt by the hand & wch in dashing upon the hand or upon white paper or a handkerchief at the distance of a quarter of an inch or half an inch from the glass or above, illuminates the hand or paper or handkerchief with a white light while the glass continues in motion, the spirit by striking upon those bodies being agitated so as to emit the light. & that if some threds of cotton or worsted yarn hanging by one end at a little distance from one another be attracted at the other end towards the glass, & a mans finger be advanced towards the attracted ends of the threds, the threds will recede from the finger, & this they will do as well when they are within the glass as when they are without it. There is therefore an electric spirit by wch bodies are in some cases attracted in others repelled & this spirit is so subtile as to pervade & pass through the solid body of glass very freely in both cases, & is capable of contraction & dilatation expanding it self to great distances from the electric body by friction. & thefore (*sic*) is elastic & susceptible of a vibrating motion like that of air whereby sounds are propagated. & this motion is exceeding quick so that the electric spirit can thereby emit light. And that wch emits light in the experiments above mentioned, may emitt it in all shining bodies whenever sufficiently agitated either by heat or by putrefaction. And the Medium wch emitts light may also be able to refract & reflect it as was noted above. This spirit may be also the Medium by whose vibrating agitations stirred up within dense bodies, the bodies receive heat & communicate it to contiguous bodies; the vibrations being propagated from one body into another where the bodies are contiguous, but reflected at the surface where they are not contiguous & by reflections kept within the hot body. The like vibrations may be excited in the bottom of the eye by light & propagated thence through the solid capillamenta of the optick nerves into the sensorium for causing vision & the like of other senses. The like vibrations may be also propagated from the brain through the solid fibres of the spinal marrow & its branches into ye muscles for agitating & expanding the liquors therein & thereby contracting the muscles to cause the motions of animals. For as (*sic*) liquors are expanded by heat & by consequence by the vibrating agitations of this spirit. If the agitations be of short continuance they expand the liquors without heating them for want of time to do it. If lasting (as in running a race, or in supporting a burden without external motion of the body) they heat the body by degrees & at length excite sweat. This spirit therefore may be the medium of sense of animal motion & by consequence of uniting the thinking soul & unthinking body.[23]

As in the later, numbered version of the Query, Newton in this remarkable document then goes on to discuss the activity of living matter. His first paragraph is relevant here, because in it he explicitly invokes the electric spirit as the source of this activity:

The vegetable life may also consist in the power of this spirit supposing that this power in substances wch have a vegetable life is stronger then in others & reaches to a greater distance from the particles. ffor as the electric vertue is invigorated by friction so it may be by some other causes. And by being stronger in the particles of living substances then in others it may preserve them from corruption & act upon the nourishment to make it of like form & vertue wth the living particles as a magnet turns iron to a magnet & fire turns its nourishment to fire & leaven turns past to leaven. ffor the living particles may propagate the vibrating motions of this spirit into into (*sic*) the contiguous particles of the nourishment & cause ye spirit in those particles to vibrate & act after ye same manner & by that action to modify the nourishment after the same manner with the living particles.[24]

There are two reasons for thinking that these passages are unconnected with those quoted previously, and were written somewhat earlier. First, their implications are rather different, because instead of being concerned with the electric spirit merely as the cause of electrical motions and the emission of light, as the others were, they assign to the spirit a very general explanatory function, much more typical of Newton's opinions at the time he composed the *Scholium Generale* than of his attitude a few years later when he wrote the "aether" Queries. More concretely, it emerges that these alternative Queries 24 and 25 were intended to go not after the final aether Query of 1717, Q. 23, but at a point indicated by Newton's note, "conjungi queant ut cohærescant. p. 340. lin. 27", which turns out to be in the middle of what is now Q. 31, but which in the Latin *Optice* was Q. 23. They were thus drafted before ever Newton drew up the aether Queries and was led thereby to change his numbering scheme. The point where the new Queries were to be inserted comes at the end of Newton's long discussion of the various short-range forces at work in the world. He apparently intended to break off the existing Query at this point and argue for the electrical origin of these forces. Only then, presumably, would he have turned as he did in the original version to wider issues such as the tendency of the world gradually to run down, were not certain forces or active principles constantly at work in it.[25]

Professor Guerlac has described another bundle of papers from this fascinating collection, in which Newton wrote out a set of "Observations" to be included in a new edition of the *Opticks*. Several of these Observations deal with electricity. As in the passage quoted above, most of the experiments are Hauksbee's, including the production of light from a rubbed globe, the prickling sensation, crackling sound and light obtained when a finger is brought near a well rubbed piece of glass, and the motion of light bodies to and from an electrified tube; and on this occasion Newton also meant to include an account of an experiment he himself had reported to the Royal Society many years earlier.[26] He at first intended his "Observations" to come after the eleven Observations already set out in Book III of his *Opticks*, and he numbered them, accordingly, from XII to XV. Later, when Desaguliers' experiments convinced him anew of the existence of an aether, he decided to add emphasis to his discussion by setting it apart as a separate Part II of Book III. He included descriptions of Desaguliers' two leading experiments (the two-thermometer experiment and a variant of the old experiment in which a guinea and a feather were let fall together in a vacuum), called these Observations I and II, and renumbered the rest as Observations III to VI. Guerlac has on very plausible grounds dated the original drafting to late 1715 or 1716, while the amendments must have been made to it after Desaguliers devised his improved "guinea and feather" experiment in 1717.

A paper of Newton's headed "De vi electrica" was published a few years ago. A reference without further qualification to "Prop. XCIV, XCV & XCV (*sic*) Lib. I" indicates that, though now located among the manuscripts relating to the *Opticks*, this was destined initially for the *Principia*. It, too, discusses the "spirit hid in all bodies, by means of which light and bodies act upon each other mutually", that same spirit which is "the cause of electrical attraction and not only reflects, refracts and inflects light but also emits it," and whose vibrations in the nerves transmit sensations and the commands of the will. "Thanks to the same spirit the particles of bodies attract each other mutually at short distances, even without friction." Some further details are also provided, and these indicate – as do the links with the *Principia* and the wide range of phenomena ascribed to the subtle matter – that the document was drafted before Newton's reacceptance of a universal aether. "This spirit is not terminated in the external parts of bodies at a mathematical surface," we are told, "but becomes rarer gradually, and the more rare part of it spreads out from bodies on all sides to short distances and gradually comes to an end." Newton also suggests that we sense different colours as a result of vibrations of different "magnitude and number" being excited in the spirit "hidden within the capillaments of the optic nerve," and he spells out again the connections he envisages between the subtle spirit and the reflections, refractions, and alternate fits of easy reflection and easy transmission of light.[27]

To summarise: we may discern in the many papers dealing with electricity from this late period of Newton's life – papers in which the influence of Hauksbee's striking experiments is everywhere apparent – two different patterns of explanation. Newton found in Hauksbee's experiments proof that ordinary "gross" matter is everywhere pervaded by an active subtle spirit, responsible for the attraction of cohesion and, when excited by friction, for the attraction called electrical. When set vibrating, it brought about the emission of light. At first Newton believed this same spirit to be responsible also for the reflection, refraction, inflection, heating effects, and alternate fits of easy transmission and easy reflection of light, not to mention the operation of the nervous system in our bodies, and perhaps other functions of living matter as well. The *Scholium Generale* dates from this period, as do the draft "De vi electrica" and the extensive new Queries 24 and 25 meant to be inserted into Q. 23 of the Latin *Optice*. Soon afterwards, Newton drew up for inclusion in Book III of the *Opticks* the set of four additional Observations incorporating Hauksbee's leading results.

Following Desaguliers' success in November 1716 with the two-thermometer experiment, Newton took up again the idea of a universally disseminated aetherial medium. It was to this that he now attributed both the workings of the nervous system and the various optical phenomena previously ascribed to the electrical spirit. He also inclined to the view that this aether was the cause of gravity. Yet he did not altogether abandon the notion of a subtle electrical spirit pervading the pores of ordinary matter. On the contrary, he continued to attribute to the action of such a spirit not only specifically electrical effects but also the emission of light from luminous bodies and, probably, cohesion. A Query concerning the cause of gravity was drafted by July 1717,[28] and it must have been soon after this that the earlier set of Observations was recast to include an account of Desaguliers' work.[29] But then Newton had second thoughts, and decided to throw his new material into the form of Queries. The two-thermometer experiment went without difficulty into a new Query, Q. 18. The discussion of electricity was at first appended to Q. 8, but in the end only the section dealing with the production of light remained there. The passages in which Newton argued again for the existence of a subtle electric spirit were first relocated so as to come immediately after the account of the two-thermometer experiment, as Q. 18B, then moved again to the end of the "aether" Queries, and eventually omitted altogether. Only a hint of Newton's real views on this subject lasted into print, namely the passage in Q. 22 quoted above.[30]

Newton's intense interest throughout this period in the phenomena of electricity and the causes underlying them is manifest from the documents quoted above. Moreover, it seems he did not hesitate to make known to his colleagues at the Royal Society his hopes in this direction. In a French work from later in the century, he is quoted on the authority of Martin Folkes (F.R.S. 1714, and later President of the Society) as having urged the Fellows more than once to continue to investigate the subject with the utmost diligence: "Mes yeux s'éteignent, mon esprit est las de travailler, c'est à vous à faire les plus grands efforts pour ne pas laisser échapper un fil qui peut vous conduire."[31]

Yet the roots of Newton's fascination with electricity during this late period of his career have remained unclear. Zev Bechler has suggested that Newton seized on it as "the fundamental force of the micro-world" in very much a *post hoc* fashion, as the only one of the three forces known to act in the macro-world – gravity, magnetism and electricity – that fitted the technical requirements of his optical theory.[32] The documents we have quoted, however, tell a different story.

To begin with, they fully confirm Guerlac's suggestion that Hauksbee's experiments inspired Newton's renewed interest in the subject.[33] They show, in particular, that Newton found in Hauksbee's work direct experimental proof that a subtle electric spirit filled the pores of ordinary matter. In fact, most of the evidence he cites derives immediately from Hauksbee.[34] The electric spirit, Newton argues, can when excited by friction be detected directly by various of our senses. It can be felt pushing against our skin when we hold an electrified rod near our face or hand, and it can be both heard and seen (as a spark, we would say) when the point of a finger approaches the rod. If we use an electrified globe instead, the spirit can once again be felt "dashing upon the hand", and the hand and other bodies are illuminated by a white light, "the spirit by striking upon those bodies being agitated so as to emit the light." Besides these new and striking effects, however, Newton takes the age-old attraction of light objects towards an electrified body as proof enough in itself of the existence of the electric spirit. In a passage which squares extremely ill with a popularly held view of his general scientific approach, he baldly asserts that "electric bodies could not act at a distance without a spirit reaching to that distance."[35]

The remainder of Newton's argument follows very naturally, and not at all after the manner suggested by Bechler. Having already decided that agitating the electric spirit led to the emission of light, Newton found it a relatively small step to suppose that a similar cause lay behind the emission of light in other circumstances, that is, that "that wch emits light in the experiments above mentioned, may emitt it in all shining bodies whenever sufficiently agitated either by heat or by putrefaction." Since the spirit can be excited in most bodies by friction, and all bodies become luminous if heated sufficiently, the spirit must be universally present in matter. And from this apparently empirically well-founded position, the very general role that Newton ascribed to the electric spirit in bringing about the phenomena of the micro-world followed by a simple application of that much-abused philosophical principle, Ockham's razor. The relevant passage is worth quoting again in its entirely:

> And if there be such an universal electric spirit in bodies, certainly it must very much influence the motions & actions of the particles of the bodies amongst one another, so that without considering it, philosophers will never be able to give an account of the Phænomena arising from those motions & actions. And so far as these phænomena may be performed by the spirit wch causes electric attraction it is unphilosophical to look for any other cause.

Even Newton's subsequent change of heart on this last point is fully consistent with the position set out here. Desaguliers' experiment, without at all undermining the earlier parts of Newton's argument, nevertheless convinced Newton of the existence of a second subtle active medium. This made the application of Ockham's razor less straightforward, and meant that Newton had to choose between the two active spirits in assigning a cause to the various optical phenomena with which he had so long been concerned. But that is all.

In Q. 31, Newton cited electricity, along with gravity and magnetism, as examples of attractive forces already known to act in the world. "These Instances," he said,

> shew the Tenor and Course of Nature, and make it not improbable but that there may be more attractive Powers than these. For Nature is very consonant and conformable to herself .... The Attractions of Gravity, Magnetism, and Electricity, reach to very sensible distances, and so have been observed by vulgar Eyes, and there may be others which reach to so small distances as hitherto escape Observation; and perhaps electrical Attraction may reach to such small distances, even without being excited by Friction.[36]

Historians have generally assumed from this that for Newton the causes of magnetism and electricity were as mysterious and unknown as he generally took that of gravity to be. I have argued elsewhere, however, that in the case of magnetism this was not so, and that we ought to take literally Newton's caveat, "What I call Attraction may be perform'd by impulse, or by some other means unknown to me. I use that Word here to signify only in general any Force by which Bodies tend towards one another, whatsoever be the Cause."[37] Though sparse, all the available evidence suggests that, far from regarding magnetism as another unexplained force acting at a distance, Newton thought he knew the cause. The evidence indicates, in fact, that he attributed the power of the magnet to circulating streams of a peculiar subtle matter that passed axially through magnetized bodies and then returned through the external air before resuming its circulation once more. This theory was not original to Newton. It derived from the similar but more complicated scheme set out in Descartes' *Principia philosophiae* (1644), and was widely accepted by leading scientists in Newton's day. Similar conclusions ought to be reached, I believe, in the case of electricity. Here, too, Newton accepted a mechanistic explanation of the phenomena at hand. Here, too, the opinions he expounded were similar to those of his leading contemporaries, and resembled those set out in Descartes' *Principia philosophiae*. Electrical attraction and repulsion were not, for Newton, examples of unexplained forces acting at a distance.

In Descartes' day, the only generally recognized electrical phenomenon was the attraction that bodies such as amber, jet, sealing wax and glass exerted, when briskly rubbed, on nearby light objects. Descartes sought to reduce this, like all other natural phenomena, to motions and impacts among particles of matter. First, he suggested that the attraction might be brought about by a process analogous to what happens when a drop of viscous liquid hangs on the end of a rod, and the rod is shaken; part of the drop is seen to stretch out, and then to return promptly to the drop. Just so, Descartes envisaged the rubbing of a body as causing little branches composed of the smallest parts of the substances to spread out into the surrounding air: as these would, he said, remain joined together, they had to return immediately to the body, and in doing so could carry along any small objects that got entangled with them.[38]

Unfortunately, this account was incompatible with what Descartes had said earlier in his book concerning the nature of glass – yet glass, too, could be readily electrified by friction. To resolve the difficulty, Descartes invoked the finest of the three kinds of matter out of which he supposed the world to be constructed, his so-called "first element." When glass was rubbed, he said, thin strips composed of this fine and very mobile matter were forced out of the glass into the surrounding air. There, however, they found no pores fit to receive them, so they returned immediately to the glass, carrying with them any small bodies with which they became entangled. Nor, Descartes asserted, was this mode of attraction peculiar to glass: on the contrary, most cases of electrical attraction were brought about by this means, not by the mechanism he had first described.[39]

To modern eyes, Descartes' two mechanisms do not seem very different. In both, the electricity of bodies (that is, their ability to attract nearby light objects) is explained in terms of the "sweeping" action of some fine matter, initially ejected from the bodies in question as a result of their being rubbed, then returning to these bodies again. To seventeenth-century authors, however, the differences were important, and highlighted what remained for almost a century one of the chief unresolved questions in electrical theory. In the mechanism that Descartes preferred, the stuff involved was a universally disseminated subtle matter, or aether, whereas in the alternative model it was the matter of the rubbed body itself which brought about the effect.

Among Descartes' successors, opinions differed as to the relative merits of these two proposals. Orthodox Cartesians such as Rohault naturally tended to be guided by their master's opinions.[40] Boyle, on the other hand, confessed himself "very inclinable" to the view that ascribed the attractions to "the Emission of the finer parts of the attrahent," and in doing so he cited Digby, Browne and Gassendi as other proponents of it (in fact crediting

Digby with being its originator).[41] A generation later, Hauksbee, too, settled for this alternative. "The Effluvia which are provok'd from the Glass," he said, "seem to be, and are nothing else but part of the same Body exerted from it by rubbing." When, soon afterwards, he discovered that glass apparently allowed free passage to the effluvia emitted by rubbed sealing wax, he was most surprised, and could only suppose that "the Figure of the Parts of Glass and Sealing-Wax, are much alike, otherwise I cannot conceive how the Effluvia of one can penetrate and pass with such ease the Body of the other, and there to act as if it was one and the same with it."

In his early papers, Hauksbee tended, like Descartes, to regard the effluvia as solid threads of fine matter. At some time prior to the publication in 1709 of his *Physico-Mechanical Experiments on Various Subjects*, however, he changed his mind and came to see them instead as making up a subtle fluid. Boyle had also seen them in this light; but whereas Boyle had regarded them as "streams" which brought about the electrical attractions by condensing and returning to the glass, Hauksbee now ascribed the attractions to the inward motion of *air*, consequent upon the "Emission and Discharge of the Electrical Matter" from the electrified body. In Hauksbee's mature theory, in other words, there were two distinct streams of fluid matter in the vicinity of an electrified body, subtle matter from the rubbed body moving outwards and air moving inwards to take its place.[42]

Newton, in a brief discussion of electricity in his famous 1675 letter to the Royal Society setting out his "Hypothesis explaining the properties of light," made clear his allegiance at that stage of his career to a pattern of explanation similar to that accepted by his contemporaries. He first described an experiment in which small pieces of paper trapped in an air space beneath a piece of glass would "be attracted and move nimbly to & fro" with all manner of motions when the upper surface of the glass was rubbed with coarse cloth. To explain this, he, like Descartes but unlike English contemporaries such as Boyle, had recourse to a universal subtle matter, and concluded from the experiment that "there is something of an æthereall Nature condens'd in bodies." This, Newton supposed, was excited by the rubbing of the glass, and spread into the surrounding air. There it activated the little pieces of paper before recondensing into the glass again:

> Now whence all these irregular motions should spring I cannot imagine, unless from some kind of subtill matter lyeing condens'd in the glass, & rarefied by rubbing as water is rarified into Vapour by heat, & in that rarefaction diffused through the Space round the glasse to a great distance, & made to move & circulate variously & accordingly to actuate the papers, till it returne into the glasse againe & be recondensed there. And as

this condensed matter by rarefaction into an aethereall wind (for by its easy penetrating & circulating through Glass I esteeme it aethereall) may cause these odd motions, & by condensing againe may cause electricall attraction with its returning to the glass to succeed in the place of what is there continually recondensed; so may the gravitating attraction of the Earth be caused by the continuall condensation of some other such like aethereall Spirit . . .[43]

Newton's extensive discussions, quoted above, from a much later period of his life reveal that so far as the cause of electrical attraction was concerned, his view had changed hardly at all during the intervening period. Contrary to what has been suggested by the Halls,[44] Newton never envisaged the "electric spirit" as a Franklin-style fluid *collected* by friction on electrifiable bodies. On the contrary, he maintained with his contemporaries that electrification consisted in the *excitation* of matter already present in bodies, and he took it for granted that the various electrical motions were brought about in a wholly mechanical way by this agitated matter. As Newton put it, the exhalation raised out of bodies by friction "carrys light bodies along with it and agitates them in various manners according to its own motions."[45] Or again,

. . . the Agent or Spirit wch glass emits by friction, is agitated with various motions like a wind, carrying along with it little light bodies by means of wch you may know wch way it moves. And since for the most part it carries those bodies towards the glass & makes them stick to it, we may thence conclude that it is the spirit by which Electric bodies perform their attraction. The friction rarefies it & makes it expand it self into much more room then while it was in the glass. But it quickly begins to condense again & in condensing & shrinking into less room it returns into the glass & in returning carries along with it little light bodies & makes them stick to the glass untill it be all condensed & returned into the glass where it was before the friction.[46]

Newton frequently referred to the cause of the electrical motion as a "spirit". This does not mean, however, that he regarded it as some kind of immaterial agency, and electricity, therefore, as an irreducibly non-mechanical force.[47] In his 1675 letter to the Royal Society, Newton used the phrases "subtill matter" and "æthereall Spirit" interchangeably. In his later writings he sometimes calls the curious substance he is discussing an "exhalation", at other times a "vapour". Both terms imply a material nature, as Newton himself makes clear elsewhere, in a draft intended for the *Principia*:

Vapours and exhalations on account of their rarity lose almost all perceptible resistance, and in the common acceptance often lose even the name of bodies and are called spirits. And yet they can be called bodies in so far as they are the effluvia of bodies and have a resistance proportional to density.[48]

Likewise, he says the electric spirit "expands", and subsequently "condenses" and "shrinks" back into the body from whence it came. It can be "agitated", whereupon it "will rush out . . . with a soft crackling noise" and "push against the finger". On occasion Newton describes its "turbulent motions", which he compares to "a wind". He also says it is "susceptible of a vibrating motion like that of air". These are all terms appropriate to a material agency, but of very dubious application to an immaterial one. The most natural reading in every case – and the number of cases adds conviction here – is the one that sees Newton's ideas about electricity as firmly set in the same "Cartesian" mould as those of most of his contemporaries. To be sure, the electric spirit is "much more subtile than common Air or Vapour"; but the difference is one merely of degree, not kind.

There remains the difficult question of what Newton had in mind when in the 1717 edition of the *Opticks* he added to the sentence, "The Attractions of Gravity, Magnetism, and Electricity, reach to very sensible distances, and so have been observed by vulgar Eyes, and there may be others which reach to so small distances as hitherto escape Observation", the remark, "and perhaps electrical Attraction may reach to such small distances, even without being excited by Friction."[49] Hawes has argued that "Newton envisaged two distinct types of electric force", and maintains that in addition to the usual friction-induced attraction over appreciable distances that is effected by the electric spirit, for Newton "it is the small particles of matter which themselves possess an electric force, and by means of their inherent electric force they are able to act upon one another at small distances."[50] This is, in my opinion, seriously to mistake his meaning. Granted, Newton on several occasions mentions a number of experiments involving what we now call cohesion and surface tension, to show that a short-range attractive force that he describes as electrical acts between the particles of matter. This does not for a moment imply, however, that he conceived of this attraction as uncaused and inherent in the particles of matter themselves. On the contrary, in the final paragraph of the 1713 *Scholium Generale*, in the document "De vi electrica", and again in the draft "Quaest 25" quoted above, he expressly attributes the attraction, for reasons discussed already, to the action of the electric spirit.

What, then, is the distinction between "electrical attraction unexcited" and the ordinary friction-induced electrical force? Both, it appears, are brought about by the immediate action of the electric spirit, but whereas in the case of ordinary electrical attraction the spirit must first be made (by the friction) to spread across the intervening space until it reaches the bodies to be attracted, in the other the particles concerned are already within the range of activity of their associated electric spirit. In other words, in the case of the short-range force, the friction is simply unnecessary!

We have seen that in the case of ordinary electrical attraction, the electric spirit acts in a wholly mechanical way to push little light bodies towards an electrified object. Newton unfortunately nowhere tells us enough about the spirit's role in bringing about short-range forces for us to reach any definite conclusions about his view of its mode of action in this case. It is possible that, at this level, he envisaged the electric spirit acting by means of certain (unexplained) short-range forces between its particles and those of ordinary matter.[51] If what I have suggested above is correct, however, and short-range attractions are really no more than ordinary electrical attractions acting in a situation where no initial friction is necessary, then every analogy would suggest that Newton probably saw these forces, too, as caused mechanically by some kind of activity, perhaps vibrations,[52] in the electric spirit.

Such a conclusion cuts sharply across the tendency of recent Newtonian scholarship to stress the centrality of unexplained forces in Newton's mature philosophy of nature. It now appears that in several important cases – magnetism, electricity, and perhaps even short-range cohesive and chemical forces – Newton remained closer to the mechanistic modes of thought of his predecessors than has been appreciated. This is so especially in respect to the actual mechanisms he envisaged as causing ordinary magnetic and electrical attractions, but also in the general similarity between the function he alloted, in the passages quoted above, to his "very subtil active vibrating spirit," and that ascribed by Descartes to his "subtle matter" or "first element". In both cases, it seems, the substance in question provided the "go" for a wide range of hidden mechanisms on whose action the workings of nature depended.

There were differences, of course, between the Cartesian "first element" and Newton's "subtle spirit". Since Descartes' universe was a plenum, his subtle matter was necessarily dense. Also, its particles could not act either on each other or on particles of other matter except through contact. By contrast, Newton described the electric exhalation as "rare", which suggests that he conceived its particles to be widely separated. He also said that it was elastic, and if his discussion in the *Principia* of the elasticity of air is any guide, he would have taken this to mean that it was composed of particles mutually repelling each other at a distance. This, certainly, is how he proposed in 1717 to explain the elasticity of his newly re-introduced aether.[53]

These differences are important. So, too, is Newton's methodological emphasis on determining the laws of action of the various forces at work in the world. Unlike Descartes, he did not concern himself primarily with discovering the underlying causes of these forces. There is no inconsistency, however, in his nevertheless taking an interest in those underlying causes, or in his adherence in a number of cases to the general pattern of explanation provided by Descartes. Here, as elsewhere, we ought not to allow our recognition of Newton's extraordinary achievements in certain branches of science to blind us to the fact that in many respects he remained a child of his time.

University of Melbourne

NOTES

* Research for this paper was assisted by a grant from the Australian Research Grants Committee.

[1] Isaac Newton, *Mathematical Principles of Natural Philosophy*, trans. Andrew Motte, rev. Florian Cajori (Berkeley/Los Angeles, 1934), p. 547.

[2] A. R. and M. B. Hall (eds.), *Unpublished Scientific Papers of Isaac Newton* (Cambridge, 1962), pp. 357, 359. See also *idem*, 'Newton's Electric Spirit – Four Oddities', *Isis* **50** (1959), 473–476.

[3] A. Koyré and I. B. Cohen, 'Newton's "Electric and Elastic Spirit"', *Isis* **51** (1960), 337.

[4] On the seventeenth-century meaning of the word "electricity", see N. H. de V. Heathcote, 'The Early Meaning of *Electricity*: Some *Pseudodoxia Epidemica*', *Annals of Science* **23** (1967), 261–275.

[5] Koyré, *From the Closed World to the Infinite Universe* (Baltimore, 1957), p. 234. Cf. A. R. and M. B. Hall, 'Newton's Electric Spirit', pp. 475–476.

[6] Isaac Newton, *Opticks, or a Treatise of the Reflections, Refractions, Inflections & Colours of Light* (New York, 1952: based on 4th ed., London, 1730), pp. 347–354. That the new edition was in print before the end of the year, and that it had already created quite a stir, is evident from the remark in a London newsletter dated 19 December 1717 that "Sir Isaac Newton has advanced something new in the last edition of his Optics, which has surprised his physical and theological disciples" (Historical Manuscripts Commission, *Report on the Manuscripts of His Grace the Duke of Portland*, V [Norwich, 1899], 550).

[7] Sometimes, however, Newton was not so careful with his choice of words. For example, in an unpublished draft, discussed in more detail below, he referred to a "subtle Aether or Aetherial elastic spirit," an "aetherial spirit or spiritual effluvium" contained within gross bodies (University Library Cambridge, Add. MS. 3970, f. 626; quoted by Henry Guerlac, 'Newton's Optical Aether,' *Notes and Records of the Royal Society of London* **22** (1967), 45–57, p. 48.

[8] Guerlac, *op. cit.*, pp. 48–52.

[9] Newton, *Opticks*, p. 348. Both Guerlac, 'Newton's Optical Aether', *passim*, and R. S. Westfall, *Force in Newton's Physics* (London/New York, 1971), p. 395, have remarked upon Newton's emphasis on explaining optical phenomena in the "aether" Queries.

[10] Newton, *Opticks*, pp. 350–352.
[11] Cf. Joan L. Hawes, 'Newton and the "Electrical Attraction Unexcited"', *Annals of Science* 24 (1968), 121–130; p. 123.
[12] Newton, *Mathematical Principles* . . . , p. 547. Cf. Koyré, *Newtonian Studies* (London, 1965), pp. 35–36, for the rendering of "fingo" as "feign" rather than "frame". J. L. Heilbron, *Electricity in the 17th and 18th Centuries* (Berkeley, 1979), p. 241, argues that Newton's "optical-gravitational aether" is actually incompatible with the earlier subtle spirit: "As the agent of cohesion, the electrical spirit should be present in bodies in proportion to their densities, while the optical-gravitational aether must stand rarer in denser bodies".
[13] University Library Cambridge, Add. MS. 3965.12, ff. 350–365, quoted by Westfall, *Force in Newton's Physics*, pp. 392–393.
[14] Desaguliers to Sir Hans Sloane, 4 March 1730/31, quoted by Guerlac, 'Newton's Optical Aether', p. 51.
[15] Newton, *Opticks*, p. 353.
[16] Newton's distinguishing between aether and electric spirits at this period has been noted by J. E. McGuire, 'Force, Active Principles, and Newton's Invisible Realm', *Ambix* 15 (1968), 154–208; pp. 179–81.
[17] University Library Cambridge, Add. MS. 3970, f. 295. This and subsequent passages from Newton's papers are published by permission of the Syndics, Cambridge University Library.
[18] Newton, *Opticks*, pp. 348–349.
[19] University Library Cambridge, Add. MS. 3970, f. 293^v. Although bearing a folio number ahead of that given to the other half of the sheet, this was evidently drafted later, because the set of dashes in "Qu 24" as written out here clearly refers back to the longer version quoted previously.
[20] *Ibid.*
[21] Newton, *Opticks*, p. 376.
[22] University Library Cambridge, Add. MS. 3970, f. 235. Cf. Westfall, *Force in Newton's Physics*, p. 416.
[23] University Library Cambridge, Add. MS. 3970, ff. 241^v – 241^r. Joan L. Hawes has quoted extracts from this document in her article 'Newton's Two Electricities', *Annals of Science* 27 (1971), 95–103; pp. 96–97. (Folios 235 and 241 are the two halves of a single folded sheet.)
[24] *Ibid.*, f. 241^r.
[25] Cf. Newton, *Opticks* (1952), pp. 397ff.
[26] University Library Cambridge, Add. MS. 3970, ff. 623–29, discussed in Guerlac, 'Newton's Optical Aether'. Cf. Newton to Henry Oldenburg, 7 December 1675; in H. W. Turnbull *et al.* (eds.), *The Correspondence of Isaac Newton* (Cambridge, 1959–77), I, 362–86; pp. 364–65.
[27] University Library Cambridge, Add. MS. 3970, ff. 427–428^v; published in Newton, *Correspondence*, V, 362–65. In the accompanying English translation (pp. 365–68), Newton seems to attribute the "fits" to vibrations in the *air*. This, however, is not the sense of the Latin original.
[28] Cf. Newton, *Opticks*, 'Advertisement II', p. [cxv].
[29] Guerlac, noting that Desaguliers first reported his improved "guinea-and-feather" experiment to the Royal Society on 24 October 1717, has argued that "unless Newton

was privy to certain preliminary experiments", he could not have recast the set of Observations, and included an account of this experiment, until then. In fact, Desaguliers did perfect the experiment somewhat earlier, and doubtless promptly communicated the result to Newton. By September, Desaguliers had his apparatus working so well that he was able to demonstrate the experiment at Hampton Court before the King and the Princess of Wales (*Philosophical Transactions* 30 [1717], 717–20). Not until October, however, did the Royal Society reassemble after its customary summer recess, and so not until then could Desaguliers formally report his results to that body.

30 An unresolved puzzle, in the light of this analysis, is Newton's decision to leave the final paragraph of the *Scholium Generale* unaltered in the third (1726) edition of the *Principia*. It was presumably soon after the appearance of the previous (1713) edition that he thought about describing the subtle spirit more explicitly as "electric and elastic". Later, it seems he envisaged deleting the entire paragraph (Koyré and Cohen (eds.), *Isaac Newton's Philosophiae Naturalis Principia Mathematica: The Third Edition (1726) with Variant Readings* [Cambridge, 1972], II, 764). Does his eventual decision not to do so indicate that towards the end of his life he changed his mind yet again about the aether, and reverted to his previous ideas about a subtle spirit associated with bodies? But if so, why did he not then include the additional adjectives describing this spirit, as Motte did shortly afterwards in his English translation?

31 Comte de Tressan, *Essai sur le fluide électrique, considéré comme agent universel* (Paris, 1786), I, xliv–xlv. Though published very late in the century, this work was in fact written much earlier, in the 1740s (cf. Bibliothèque Nationale Paris, MS fr. 12280).

32 Bechler, 'Newton's Law of Forces which are Inversely as the Mass: a Suggested Interpretation of His Later Efforts to Normalise a Mechanistic Model of Optical Dispersion', *Centaurus* 18 (1973–74), 184–222; pp. 215–16.

33 Guerlac, 'Francis Hauksbee: expérimentateur au profit de Newton', *Archives internationales d'histoire des sciences* 16 (1963), 113–28; *idem*, 'Sir Isaac and the Ingenious Mr Hauksbee', pp. 228–53 in I. Bernard Cohen and René Taton (eds.), *Mélanges Alexandre Koyré: Tome I, L'aventure de la science* (Paris, 1964); *idem*, 'Newton's Optical Aether'.

34 Hauksbee's results were first published in numerous articles in the *Philosophical Transactions* in the years 1705–11, and were subsequently brought together in his *Physico-Mechanical Experiments on Various Subjects* (London, 1709; 2nd enlarged ed., 1719). I have discussed his work in my paper, 'Francis Hauksbee's Theory of Electricity', *Archive for History of Exact Sciences* 4 (1967), 203–17. Gad Freudenthal points out ('Early Electricity between Chemistry and Physics: The Simultaneous Itineraries of Francis Hauksbee, Samuel Wall, and Pierre Polinière', *Historical Studies in the Physical Sciences* 11 (1981), 203–229; p. 215) that Newton also drew upon the experiments of Dr Samuel Wall.

35 See above, p. 198.

36 Newton, *Opticks*, p. 376.

37 *Ibid.* Cf. R. W. Home, '"Newtonianism" and the Theory of the Magnet', *History of Science* 15 (1977), 252–66.

38 Descartes, *Principia philosophiae* (1644), in C. Adam and P. Tannery (eds.), *Oeuvres de Descartes* (Paris, 1897–1913), VIII, 311.

39 *Ibid.*, pp. 312–314.

40 Jacques Rohault, *A System of Natural Philosophy* (London, 1723; reprinted New York, 1969), II, 186–187.

[41] Boyle, *Experiments and Notes about the Mechanical Origine or Production of Electricity* (London, 1675: reprinted Oxford, 1927), pp. 3–6.
[42] R. W. Home, 'Francis Hauksbee's Theory of Electricity', *passim.*
[43] Newton, *Correspondence*, I, 365.
[44] A. R. and M. B. Hall (eds.), *Unpublished Scientific Papers of Isaac Newton*, p. 209.
[45] University Library Cambridge, Add. MS. 3970, f. 293^{v}.
[46] *Ibid.*, f. 628. Newton's failure to concern himself here with explaining electrostatic repulsion suggests that, like those of his predecessors who had noticed it, he regarded the effect as a mere mechanical rebounding and not a real phenomenon. Hauksbee thought otherwise, for reasons of which Newton must have been well aware.
[47] Cf. McGuire, 'Force, Active Principles, and Newton's Invisible Realm', p. 176, where Newton's assertion, "And so far as these phænomena may be performed by the spirit wch causes electric attraction it is unphilosophical to look for any other cause" (cf. p. 198 above) is taken to mean that " 'it is unphilosophical to look for any other cause' of electrical attraction." As indicated earlier, I, on the contrary, see the passage as an invocation of Ockham's razor with respect to the cause of "the motions & actions of the particles of the bodies amongst one another" more generally. I do not see it as an abandonment of the search for a mechanical explanation of electricity.
[48] University Library Cambridge, Add. MS. 3965.13, f. 437^{v}; quoted by McGuire, 'Body and Void in Newton's De Mundi Systemate: Some New Sources', *Archive for History of Exact Sciences* 3 (1966), 206–48, p. 219 (Latin original p. 245).
[49] Newton, *Opticks*, p. 376.
[50] Hawes, 'Newton and the "Electrical Attraction Unexcited" ', pp. 125–26; and *idem*, 'Newton's Two Electricities', *passim.*
[51] Heilbron, Note 12 above, p. 240.
[52] See above, p. 198. Another document tells us that the "agitation" of the spirit *softens* hard bodies and converts them into liquids (University Library Cambridge, Add. MS. 3970, f. 604). The same document adds chemical forces to the range of phenomena explained in terms of the electric spirit. Relevant passages are quoted, and an English translation provided, in McGuire, 'Force, Active Principles, and Newton's Invisible Realm', pp. 177–78. Since elsewhere in this document Newton attributes reflection, refraction and so on to the electric spirit, the whole presumably pre-dates his return to the idea of a genuine aether in late 1716.
[53] Cf. Newton, *Mathematical Principles* . . . , pp. 295–302, and *Opticks*, p. 352.

G. A. J. ROGERS

THE SYSTEM OF LOCKE AND NEWTON*

INTRODUCTION

There is wide agreement that both Locke and Newton are co-founders of the Enlightenment. The question therefore arises as to why it was that they succeeded in supplying such a long-lived, and influentially respectable, foundation for much of eighteenth century and later thought. It is to this, undoubtedly large, question that this paper is a contribution. My major task will not, however, relate to the way in which Locke and Newton appealed to their followers. It will not, in that sense, be primarily a contribution to the social history of intellectual ideas. Rather I shall try to show the coherence of their views and the overlap, indeed often the intimacy, between their positions. In this way it will be possible to see that their philosophies supplied a system, an intellectual system in the grand sense in which Aristotle, Aquinas, Descartes, Kant, and half a dozen other thinkers have provided such systems.

Although the major achievements in science and philosophy of Newton and Locke have long been known, our modern categories of study have tended to obscure their shared achievements. The historians of science have claimed Newton for their own. Locke, not widely recognised within the history of science, has been left to the historians of philosophy, who themselves have often been less concerned with history than with philosophy.[1]

This separation of Locke and Newton has not always been the accepted approach. Some thinkers in the eighteenth century, for example, recognised the unity of their thought. But in important respects that unity was masked by the fact that there was little overlap in their printed work. Thus, although Voltaire recognised the contiguity of their positions, he was unable to see how close that fit was. Indeed, it is only with a study of the unpublished papers of the two men that its full extent may be appreciated. As an illustration we might note that the manuscripts of Locke reveal a man who was extensively involved with the science of his age, whilst Newton's papers show that he often pondered problems both metaphysical and epistemological.[2]

The form of my argument will be this. I shall briefly outline the intellectual relationship between Locke and Newton. I shall then argue that by the early eighteenth century at the latest – the *impact* of the *Essay* was in relevant

Z. Bechler (ed.), Contemporary Newtonian Research, 215–238.

circles quite rapid – the *Essay* was widely accepted in England as the best account of human understanding available. I shall then suggest that Newton both read and generally approved of Locke's *Essay* – as Locke most certainly did of the *Principia* – and that Newton in some important sense became a Lockian. I shall argue that corroborative evidence for my claim is to be found in the writings of the Newtonians, those expositors and popularisers of Newtonian science of the early and middle eighteenth century. Finally I shall suggest a few of the reasons why it is appropriate to see the work of Locke and Newton as a system.

THE RELATIONSHIP BETWEEN LOCKE AND NEWTON[3]

The widely accepted view of the relationship between Locke and Newton is one which sees Locke's whole intellectual position as indebted to the Newtonian achievement, whilst it is assumed that Newton learned little, if anything, from Locke.[4] As against this, I have argued that there was no major issue on which Locke felt himself required to review his position in the light of Newton's work. The *Essay Concerning Human Understanding* was virtually complete in its first edition when Locke first read the *Principia* and he felt no need either to revise the text or make substantial changes in subsequent editions because of it. It is also relevant to remember that Locke saw only the first edition of Newton's work, for he died in 1704, nine years before the second edition appeared. Newton's *Opticks* could have no influence, for it was first published only months before Locke died, although no doubt Locke was well aware of Newton's optical work, either directly from Newton or from his numerous contacts among members of the Royal Society. In short, I believe that we may say with confidence that rather than the *Essay* being a product of Locke's absorbtion of Newton's methodology, as is sometimes suggested, Locke found in Newton's work an exemplification of principles of method in the natural sciences to which he was already totally committed.

With regard to any possible influence of Locke on Newton, again, contrary to received opinion, there is now real evidence that Newton may well have found in Locke's *Essay* a statement of an epistemology which not only coincided with his own beliefs, but which penetrated further into the issues than Newton had himself ever done. In other words, to a substantial extent Locke supplied the detailed epistemology to accompany the *Principia* and Newton probably borrowed from that epistemology to support his own position. A summary of the evidence for these claims will be given below.

Directly relevant to an understanding of the relationship between Locke and Newton is the extent to which each was at home in the major areas of intellectual enquiry of the other. Let us, then, first take notice of Locke's involvement with the natural sciences.

Locke had gone up to Christ Church in 1652 at a time when science was beginning to flourish at Oxford as it had never done before, and he was soon involved in the new movement. His interests seem to have begun with medicine, and from his notebooks we may see that even as an undergraduate he became a keen student of experimental physiology. He attended the lectures of Thomas Willis, taking full and careful notes. He was at this time seriously thinking of becoming a professional physician, and it was not long before he was collaborating with Richard Lower in some of the latter's most productive research. Later, Locke was to be the assistant to an even more distinguished physician, Thomas Sydenham, and to collaborate with Robert Boyle in his researches on human blood.[5]

Locke's notebooks also reveal that he was a diligent student of chemistry.[6] He attended the chemistry classes arranged by Robert Boyle and given to a distinguished audience by Peter Stahl. Soon after the death of Boyle chemistry was the subject of an exchange of letters between Locke and Newton.[7]

Although Locke may be regarded as highly proficient in medicine and well-grounded in chemistry, there is undoubtedly a question mark over his mastery of mathematics and physics, even though most commentators have exaggerated the evidence which is held to support the view that he was a poor mathematician.[8] If it is true that Locke could not follow all of the proofs of the *Principia* this would not in itself show him to be innumerate, for many professional mathematicians were in the same position.[9] If he mastered the mathematical texts which he was recommended to buy then he must have reached a fairly high level of comprehension and competence.[10] And if he both attended and understood the lectures on geometry and astronomy which his course at Oxford required then this too would suggest that it would be wrong to assume a general ignorance on such matters. Granted this, therefore, there is no reason to doubt, and much reason to suppose, that with this sort of background, there were many matters in the area of the natural sciences, to say no more, which Locke and Newton might well wish to discuss when they became acquainted on Locke's return from Holland in 1689.

What, then, of Newton's concern with matters of philosophy, more especially those of epistemology, ontology, and metaphysics? Here we find Newton's manuscripts particularly fruitful. There is, to begin with, the sections of the early notebook, containing the famous *Questiones quaedam*

Philosophicae, which not only includes many matters relating to natural philosophy but also has many entries directly concerned with central areas of epistemology and metaphysics. It is clear from this notebook that Newton was concerned with a whole series of such questions. Many of them were to receive detailed answers in Locke's *Essay* over two decades later. A full analysis of this notebook is much needed for a proper understanding of Newton's early intellectual position.[12] In particular, whilst historians of science have examined many of the scientific positions contained therein, as yet there has been comparatively little attention given to Newton's philosophical ideas, often borrowed, and no doubt far from always settled at this early stage of his intellectual career.

We shall be turning to consider some specific entries at a later stage, but for the moment let us just note that the existence of the work testifies to Newton's concern with such questions. Nor, for several reasons, is this surprising. The Cambridge of Newton's student days was one vibrating with new ideas, and many of these were philosophical in nature. The writings of Gassendi and Descartes were becoming familiar, presenting major challenges to the orthodox undergraduate courses. And Cambridge had its own important groups of thinkers, among whom Henry More and other Platonists were prominent. Newton's notebook reflects that invigorating atmosphere.[13] Furthermore, Newton's own tutor, Isaac Barrow, was clearly a man who gave considerable attention to matters of methodology, to the philosophy of science, as we might say today.[14] And Newton's early engagement with issues of methodology probably owes much to him.

There is a more general consideration, too, which was likely to make Newton keenly interested in questions of philosophy. It was that questions about the foundations of knowledge were very much in the air and the scientist saw himself as duty bound to offer some general account of the nature of knowledge and the proper method of its acquisition in his scientific works. Thus Walter Charleton, following Gassendi, makes such an offering at the beginning of his *Physiologia Epicuro-Gassendo-Charltoniana*. It is of some interest, therefore, to find the *Questiones* headed with the words '*Amicus Plato, amicus Aristoteles, magis amica veritas*', and to discover that these words are to be found near the beginning of Charleton's work.[15]

Another dimension is given to the importance of philosophical questions for the young Newton when we reflect that at almost every turn there were both religious and scientific implications for the answers offered to such questions as those of the nature of substance, of perception, of God, or of the Soul.

A real problem with this early notebook is to know how far it may be taken as a reliable guide to Newton's own beliefs. Since many of the entries are quotations, and since to note another's views is not necessarily to subscribe to them, there are clearly major difficulties of interpretation. However, as many of the positions stated re-emerge in Newton's mature thoughts, there are at least some fairly strong inferences about his beliefs at this time which may be drawn.

Although there are several intermediate sources for Newton's philosophical ideas, the most important other manuscript sources post-date Newton's first contact with Locke in about 1690, and many more post-date Locke's death in 1704. Centrally important are the drafts for the *Regulae Philosophandi* and the successive drafts of the *Queries* to the *Opticks*. It is these manuscript sources especially, that reveal how close the intellectual positions of Locke and Newton were, and which also reveal that Newton was concerned to place his scientific ideas on a sound philosophical basis, even if he was not prepared to argue that basis in detail in public. What we can surely say, on the basis of the manuscript sources, is that Newton was concerned with questions of philosophy and that the common ground between him and Locke on philosophical issues was very substantial, not only in general, but also in detail.

The exact date on which Locke and Newton met is not known. Although it is possible that they were acquainted before Locke went to Holland in 1683,[16] there is no firm evidence and the traditional view that they met at the salon of the Earl of Pembroke in late 1689 or early 1690, soon after Locke's return from Holland, is probably correct. By this time each was the author of his most famous work and both were central figures in the intellectual landscape of the Royal Society, and known supporters of the Whig Revolution.

From 1690 onwards we have a record of many key aspects of the friendship which was rapidly established between the two men. It is a relationship which is not difficult to explain, granted their shared interests, knowledge, and the high intelligence which each could bring to bear on issues of mutual concern. Both were keenly interested in the natural sciences, though Locke's interests were wider and shallower than Newton's, both were deeply religious and believed that religion was capable of rational comprehension. Their family backgrounds were far from dissimilar. Both came from remote rural areas and from modest middle class homes. Each remained a life-long bachelor, both were disinclined to rush into print. More specifically, their intellectual outlooks were remarkably alike. Whig in politics, they were strongly opposed to enthusiasm in matters of religion. Both, whilst strongly attracted to the philosophy of Descartes, were in fact set against what they took to be the fundamental flaws both in Cartesian theory and in Cartesian methodology. Each was committed to the method in science which found expression in the work of the Royal Society. It is not too much of an exaggeration to say that a detailed account of the shared intellectual attitudes of Locke and Newton would go a long way to describing most of the major intellectual forces which shaped the eighteenth century. We shall explore implications of some of these links below. But for the moment just let it be acknowledged that there was sufficient in common between the two to make for a fruitful relationship. To be sure they had their moments of difficulty. Particularly so when Newton suffered from a severe mental disturbance in 1693, but, through Locke's tact, no permanent damage was done to their friendship.

THE RECEPTION OF LOCKE'S ESSAY

When Locke's *Essay* was published in 1690 it was not universally welcomed. (Newton himself called Locke a Hobbist, though he was later embarrassed by his remark).[17] But it was amongst the fellows and fellow-travellers of the Royal Society that the *Essay* received its warmest welcome. It was dedicated to the then President of the Royal Society, Lord Pembroke, who had strongly encouraged Locke in his publishing enterprise. And it can in many respects be regarded as an expression by the most able philosopher of the Royal Society group of their contemporary view of human understanding. John Aubrey described the *Essay* as the "best Booke that ever was writt by one man".[18] And Evelyn wrote to Pepys (in 1690) in terms that can be judged fulsome in the extreme even by seventeenth century standards. Locke, he wrote, had

> baited and entertained me with so rare and extraordinary a piece that 'twas not possible for me to dismiss him at one reading . . . there has nothing appeared . . . written with so much force, upon the subject which he handles . . . I look upon it as what may serve for Institution, as well as Instruction in the most necessary the least understood part of real philosophie as far as our attainements can pretend in this Unenlightened state.[19]

These responses should also be seen in terms of the way in which Locke's philosophy became incorporated into the official accounts of knowledge, especially scientific knowledge, which poured from the press in the early eighteenth century. One such was John Harris's *Lexicon Technicum* in which the entries under *Ideas, Identity, Knowledge*, are all straight Locke. (See especially the 5th edition of 1736). And in the many various popular expositions of Newton's science we constantly find the language and thought of

Locke. Thus, there is more than an echo of Locke in Henry Pemberton's *A View of Sir Isaac Newton's Philosophy* (London, 1728) when he writes in the Introduction:

The proofs in natural philosophy cannot be so absolutely conclusive as in mathematics. For the subjects of that science are purely the ideas of our own minds. They may be represented to our senses by material objects, but they are themselves the arbitrary productions of our own thoughts; so, that as the mind can have a full and adequate knowledge of its own ideas, the reasoning in geometry can be rendered perfect. But in natural knowledge the subject of our contemplation is without us, and notso compleatly to be known: therefore our method of arguing must fall a little short of absolute perfection. It is only here required to steer a just course between the conjectural method of proceeding, against which I have so largely spoke; and demanding so rigorous a proof, as will reduce all philosophy to mere scepticism, and exclude all prospect of making any progress in the knowledge of nature.

Strong Lockian themes are to be found in other Newtonian expositions. s'Gravesande's *Mathematical Elements of Natural Philosophy confirm'd by Experiments: or An Introduction to Sir Isaac Newton's Philosophy* contains an entirely Lockian epistemology. He writes, for example:

Being about to deliver the Nature of mathematical Evidence, and to tell why no one in his Senses, can deny his Assent to it, I shall consider the Nature of our Mind.

Our Mind contains Ideas, and compares, them one with another; in this all Knowledge is placed.

Our Mind perceives Ideas, is conscious of this Perception; and no one can doubt, whether it truly perceives the Idea, which it perceives.

I don't here speak of the Agreement between these Ideas and external Things, I only speak of that Representation which is present to the Mind. While the Idea of a Building is present to my Mind, I may make it a question, whether there is really any particular Building without me answering to this Idea, or can be; but I shall not doubt of this viz. whether I think of a Building which is represented to my Mind for the Mind is conscious of its own Perception.

The Mind is not always sensible of one only Idea at a time; but the greater its Capacity is, the more Ideas it takes in at a time.

But, when many Ideas are together present to the Mind, the Mind necessarily perceives the Comparison of them, and forms an Idea of this Comparison to itself.[20]

And later he gives Locke's definition of knowledge:

You have that, by which you may solve the Difficulties, in which Scepticks endeavour to involve the Truth. Let them say that Truth stands in need of a Criterium, that a true Criterium is to be distinguish'd from a false one, and that a new Criterium is wanting for this Purpose; which new Criterium will likewise want its Criterium, and so on in infinitum: therefore, that it is a Contradiction to suppose any Criterium.

> I answer, that the Evidence itself is the defin'd Criterium of Truth; viz. the very Perception of the Comparison between two Ideas. That the Criterium of Evidence is Consciousness; but that this brings its own Criterium along with it: for I do not want a Criterium to make me certain, that I am conscious of the Idea that is present to my Mind. Can I avoid perceiving the Idea which I do perceive? While I am conscious, am I therefore not conscious that I am conscious? To demand another Criterium of Consciousness implies a Contradiction: but I have just prov'd, that this is the only Foundation of mathematical Evidence.[21]

These examples are, I believe, sufficient to establish that Locke's epistemology was intimately linked with the popular expositions of Newton's views.

Why, then, did these authors think it necessary to include the Lockian epistemology in the Newtonian work? The answer must surely be that for one reason or another they thought that the epistemological premises of Newton's philosophy needed to be made clear and that they found in Locke's work a clear statement of the philosophical basis for the method actually employed, at least as they read it, in the *Principia*. In other words, they read the *Principia* as an exemplification of the method to science which Locke had argued was the only possible route to secure knowledge of the natural world: the *Essay* was taken to supply the philosophical bedrock of Newtonian science.

To have borrowed from Locke so easily, one might say almost unnoticed by modern eyes, reveals how close the match was between the *Essay* and the *Principia*. But it probably reflects another aspect of the whole context of the presentation of Newtonian science. Newton's work had to be defended against the charge that, unlike its continental rival, the Cartesian system, it was insecurely based in metaphysics. The *Essay* was seen as fulfilling for Newton's physics the role that the *Discourse on Method* played for the science of Descartes.

Such an interpretation of the seen relationship between the *Essay* and the *Principia* does not require us either to suppose that either work was directly indebted to the other for its production. And, as we have seen, no such debts exist. But once produced, and produced so adjacent, they were admirably placed to support each other. Locke appears to have grasped the point immediately. Hence, we may surmise, the famous reference to the "incomparable Mr. Newton" in the Epistle to the Reader of the *Essay*. But what of Newton? Did he in any sense see the *Essay* as providing the philosophical foundations upon which his science might rest? There is no certain and simple answer to that question, but the evidence suggests an affirmative answer: Newton found in Locke's *Essay* a sustained analysis and justification for an epistemology and ontology to which he was already committed, but for which he never himself supplied the detailed argument. We must now turn to the evidence to support that claim.

THE EPISTEMOLOGIES OF LOCKE AND NEWTON[23]

Is there, then, any evidence in Newton's writings to support the view that in some real sense Newton was a Lockian? Here we find the manuscripts of Newton more revealing than the printed age. Let us begin with the important draft Rule V of the *Regulae Philosophandi* to which Koyré and Cohen have already drawn attention.[24] The draft, written about 1714, ten years after Locke's death, reads as follows:

> Whatever is not derived from things themselves, whether by the external senses or by the sensations of internal thoughts, is to be taken for a hypothesis. Thus I perceive that I am thinking, which could not happen unless at the same time I were to perceive that I exist. But I do not perceive that any idea whatever may be innate. And I do not take for a phenomenon only that which is made known to us by the five external senses, but also that which we contemplate in our minds when thinking: such as, I exist, I believe, I understand, I remember, I think, I wish, I am unwilling, I am thirsty, I am hungry, I rejoice, I suffer, etc. And those things which neither can be demonstrated from the phenomenon nor follow from it by the argument of induction, I hold as hypotheses.[25]

I have elsewhere made comparisons with the various claims made by Newton in this Rule and the text of Locke's *Essay*, and I have claimed that similarities revealed cannot be lightly dismissed.[26] A remarkable number of further resemblances are to be found in a comparison between Newton's scattered remarks on epistemology and ontology and the considered opinions of Locke as expressed primarily in his published works, and especially the *Essay*. To be sure there are some differences, but in general these are as nothing (except in one important area) compared with their similarities. The one area of exception is that relating to their views on space and time.[27] The general wealth of resemblance strongly suggests that Locke's work added considerably to the philosophical depth of Newton's own, largely empiricist, epistemology.

Before we begin such comparisons a word of caution is in order. Locke wrote a major work of epistemology and Newton did not. Further, whilst Locke is remembered primarily as a philosopher, Newton equally obviously is remembered primarily as a mathematical physicist. As a result, whilst the text of the *Essay* is often taken as sufficient evidence for Locke's position on a matter of epistemology, Newton's position is not so clear. Are we to accept as Newton's beliefs what Newton says? Or should we be only impressed by his actual working practice as a scientist? The matter is of some importance as there is mounting evidence that there is often a disparity between Newton's claims as to method and his actual practice.[28]

To help us over this difficulty I shall employ the following distinction. I shall label it as that between the logic of justification and the logic of discovery. Whether there is a *logic* of discovery is doubtful, though clearly some techniques are more likely (have in the past been more fruitful) in bringing about discoveries than others. But there is surely, in some minimal sense at least, a logic of justification. There is, that is, a variety of possible grounds for somebody believing that an explanation of natural phenomena (to take no more) is true or probably true, and the relationship between evidence and conclusion may be rationally assessed. Newton believed that only in certain situations, i.e. only when certain conditions were satisfied, did the case for the acceptance of some statement about the natural world demand our assent. For Newton these grounds were largely to be understood ultimately in terms of direct empirical evidence for the truth of the statement.

But this is far from claiming that Newton believed only those things for which he had such direct evidence. Often, like the rest of us, he was convinced of the truth of something for which, on his own account, he did not have sufficient evidence.[29] What does appear to be true, however, is that Newton did not believe that such ill-supported conjectures should be allowed to be propagated as pieces of natural philosophy. At least in principle Newton was an inductivist.

Clearly Locke and Newton shared similar ontologies. The influence of Descartes on each, either directly or indirectly, was substantial.[30] Like him, they were dualists, believing in mind and matter as two separate substances. Each subscribed to a causal theory of perception, and each was ambivalent about the objects of perception, switching from the language of ideas to that of some form of direct realism. Each claimed that the origin of ideas was experience, and denied the existence of any innate ideas. Both of them clearly believe the physical world to be one in which there occur genuine causal interactions between physical objects. Both of them thought the Scholastic account of substance and properties to be totally misconceived and, whilst both believed in some form of corpuscular theory of matter, each acknowledged that theories of matter were not as yet well grounded. Each recognised the obscurity of the notion of substance. They agreed that general propositions about the world were not logically certain. Such propositions should be based upon observation, and were liable to exception. It followed from this that both rejected any aspiration to an *a priori* physics. On the other side, each totally rejected the view that all that could be expected was some sort of hypothetical physics. Each accepted that the properties of matter could be divided into two sorts: the famous primary-secondary quality distinction. Each believed in God, and believed that God's existence was evidenced by experience, though their proffered proofs differed.

Such a list points to a substantial body of intellectual agreement between the two men. But these agreed points should not cause us to ignore important differences. Some of these differences include the following: although their ontologies were much the same, they were not identical. Newton's list of primary and secondary qualities, for example, is different from that of Locke. This in part reflects Newton's much more sophisticated account of the origin and nature of force, a central achievement of his physics and his philosophy of nature. Put at its simplest, Newton seems to have come to believe that matter only possesses passive qualities, and that active properties were probably the direct intervention of God in the physical world. Locke was more cautious on these issues and there is no evidence that he gave the matter the attention which Newton did.[31]

But if Locke has left us little to suggest that he devoted much time to speculative thought about the ultimate forces of nature, Newton offers us no detailed argument for much of his epistemology. Here the two men were in important respects complementary. It is precisely in this sense that they were the twin founders of much of eighteenth century thought.

There are other points of difference. Locke, as we have already noted, was not a great mathematician even though he saw the relevance and power of mathematics to handle many problems. Newton was always the mathematician, seeking ways to embrace phenomena within the quantified fold. But Locke's mathematical weaknesses, if they may be so called, must be seen in the light of an undoubted strength, his substantial contributions to our understanding of the nature of language. Here Newton shows no such comparable quality. A major contribution of Locke is his rejection of the possibility of any sort of definitive scientific language, a *characteristica universalis*. Newton at an early stage, seems to have been attracted by the project of John Wilkins and others to produce such a language, a project upon which Locke's *Essay* might be seen as turning the final screw.[32]

Although, as we have already hinted, both Locke and Newton were opposed to speculative theory in science, there were some differences between their two views. Despite Newton's well-known hostility to hypotheses in natural science, he was, in the end, much more prone to them – if that is the right expression – than Locke was. In part this reflects the areas of natural science with which each was most concerned. Seventeenth century mathematical physics probably gained more from speculation than did either seventeenth century medicine or chemistry, the two branches of science in which, as we have seen, Locke was most deeply engaged. Probably, also, it is a measure of their innate interest in the sciences. Newton's mind was for ever probing the secrets of nature, Locke's was only intermittently so.

Locke's well-known rejection of innate ideas is the negative half of his positive thesis that all the ideas with which the mind is furnished are the product of experience, through the two routes of sensation and reflection.[33] Locke was well aware of important implications of his claim. One of these related to moral theory and another to the natural sciences. The latter – rejection of innate speculative principles – is the one that now concerns us.

Although his attack on such principles was undoubtedly aimed widely it certainly included the Cartesians. In the first Book of the *Essay* Locke makes a special point of denying that we have an innate idea of substance,[34] and in Book Two, in a sustained attack on Cartesian theory he says:

> To ask, at what time a Man has first any ideas, is to ask, when he begins to perceive; having ideas, and perception being the same thing. I know it is an opinion, that the soul always thinks, and that it has the actual perception of ideas in itself constantly, as long as it exists; and that actual thinking is as inseparable from the soul, as actual extension is from the body; which if true, to enquire after the beginning of a man's ideas, is the same, as to enquire after the beginning of his soul. For by this account, soul and its ideas, as body and its extension, will begin to exist both at the same time.[35]

To believe that the soul always thinks "is to be sure, without proofs, and to know without perceiving", itself, in Locke's eyes, a contradiction. " 'Tis . . . a confused notion, taken up to serve an hypothesis . . . "[36] which was to accuse the Cartesians of building upon a fundamentally unsound method in philosophy, reinforced a little later by

> He that will suffer himself, to be informed by observation and experience, and not make his own hypothesis the rule of nature, will find few signs of a soul accustomed to much thinking in a new born child, and much fewer of any reasoning at all.[37]

Newton's philosophical views about the origins and nature of knowledge were never fully expressed in his published works. But in his manuscripts there are some important sources. The single most relevant text is the draft Rule V to which we have already referred. As we have seen his position was entirely Lockian. The rejection of innate ideas, the distinction between ideas of sensation and reflection, and the attack on hypothetical explanation are all common themes. Was it Newton's considered opinion? The evidence suggests that it was. There are several other manuscript sources stating the same or similar positions. Further, the view expressed in the draft Rule is entirely consistent with all his published methodological and epistemological remarks, and, so far as I have discovered, all his unpublished ones as well. But Newton, unlike Locke, offers no sustained argument for his views. Perhaps he believed no detailed argument was called for after the publication of Locke's *Essay*. It also suggests that Newton did not pay a great deal of attention to matters of epistemology. Just as Locke by temperament was not strongly attracted to the mathematical sciences, so Newton was only weakly moved towards epistemology.

Granted the shared rejection of innate ideas as a source of knowledge, what are the comparisons which may be made between their positive accounts of the origins of knowledge in perception? Here a detailed comparison reveals how close their positions were.[38] Both were clearly dualists, in the sense that each held that it was the mind or soul which saw, and what it saw was the product of a causal chain extending from the external world via the senses to the brain. Newton expressed his position clearly in a letter which predates any known contact with Locke, thus:

> Let us now consider what may be the cause of the various situations of things to the eyes. If when we look with one eye it be asked why objects appear thus and thus situated one to another, the answer would be because they are really so situated among themselves and make their collored pictures in the retina so situated one to another as they are; and those pictures transmit motional pictures into the sensorium in the same situation, and by the situation of those motional pictures one to another the soul judges of things without.[39]

Both Locke and Newton believed that the empirical evidence dictated such accounts, and no doubt both were aware (Locke's awareness found clear expression in the *Essay*) that there was an unsolved problem as to how it was that motions in the brain might result in ideas in the mind, underlining the way in which the issue of dualism was intimately connected with accounts of perception. Both, too, were ambivalent over the problem as to what exactly the objects of perception are. Do we see ideas, or such things as chairs and tables? What exactly the phenomena are, ontologically speaking, receives from each of them two answers in different places. They, like others before and since, were unable to offer a satisfactory solution to this central problem of empiricist epistemology.

For both Locke and Newton their position with respect to dualism was closely, and unsurprisingly, linked to the problem of substance. Some of the issues involved, so far as Newton was concerned, surface in his puzzlement over double vision, discussions of which occur several times in his papers.[40] The roots of this puzzlement appear in the early notebook to which we have already referred, in passages which probably owe more than a little to Henry More, and which are directed against Hobbes's materialism. Newton, in these immature reflections – though many were to become his considered opinions – links double vision, or rather the lack of it in normal circumstances, with his dualism. His argument is that if the soul were "nothing but modified matter and did memory consist in action" we could never recall things at will. Our memories would depend only upon the motions in the brain. Further, "If sense consisted in reaction we should perceive things double . . . ".[41]

Two things which emerge clearly from Newton's remarks in this notebook are that he believed empirical evidence – the phenomena – point unambiguously towards dualism, and that it is only from the phenomena that we may learn more about both soul and body. Under the heading 'Philosophy' he entered:

The nature of things is more securely & naturally deduced from their operacions one upon another y^n upon oer senses. And when by ye former experiments we have found ye nature of bodys, by ye latter wee may more clearely find ye nature of o^r senses. But so long as wee are ignorant of ye nature of both soule & body wee cannot clearely distinguish how far an act of sensation proceeds from ye soule & how far from ye body Etc.[42]

In a series of questions and answers entered under 'Of ye soule' Newton's dualism emerges unambiguously and powerfully:

Memory is a faculty of ye soule (in some measure) for else how can divers sounds, or words excite her to divers thoughts or 3.4.5 or more words beget ye same thought in her. Perhaps she remembers by ye helpe of characters in ye Braine, but yn how doth shee remember the signification of those characters.

Quase 1. why Objects appeare not inversed Respt. The mind or Soule cannot judge ye image in ye Braine to be inversed in less she perceived externall things wth wch shee might compare y^t Image.

2. Why doe [objects?] appeare to bee without [i.e. outside] our body? Resp: Because in the image of things delineated in the braine by sight, ye bodys image is placed ye midst of ye images of other things is moved at o^r command towars & from those other images, &c.

3. But why are not these objects then judged to bee in the braine Resp. Because ye image of ye braine is not painted there, nor is ye Braine perceived (*sic*) by ye soule it not being in motio, & probably ye soule perceives noe body but by ye helpe of their motion. But were ye Braine perceived together with those images in it wee should thinke wee saw a body like the braine encompassing & comprehending our selves ye starrs & all other visible objects. &c.[43]

Newton clearly believed on both direct empirical grounds, and on the basis of inference, that we have both a body and a soul. We can be sure also, both on independent grounds and from the implications of his remarks, that he thought the soul to be an immaterial entity. Matter was governed by the laws of motion (still far from understood by Newton in the 1660's when he made his notebook entries). It was essentially passive. Mind, however, was active, capable of being a direct and sufficient causal agent, as exemplified in deliberate human actions, whether mental or physical: "Were ye soule nothing but modified matter . . . we could never call things into o^{r} memory."[44]

But what of the real nature of both soul and body? Here Newton, like Locke, and in sharp contrast to Descartes, is very cautious and uncertain. Of Newton's conception of the nature of the soul there is almost no evidence. There is much more, however, on the nature of body. Much more than can be satisfactorily discussed here. I shall therefore confine my remarks to some aspects of Newton's account which are particularly relevant to a comparison with Locke.

In a comparatively early composition *De Gravitatione et aequipondio fluidorum* of about 1671,[45] Newton puts forward a rich collection of ideas about matter, many of which were to remain with him until the end of his life. His account of body begins as follows:

> Now that extension has been described, it remains to give an explanation of the nature of body. Of this, however, the explanation must be more uncertain, for it does not exist necessarily but by divine will, because it is hardly given to us to know the limits of the divine power, that is to say whether matter could be created in one way only, or whether there are several ways by which different beings similar to bodies could be produced. And although it scarcely seems credible that God could create beings similar to bodies which display all their actions and exhibit all their phenomena and yet are not in essential and metaphysical constitution bodies; as I have no clear and distinct perception of this matter I should not dare to affirm the contrary, and hence I am reluctant to say positively what the nature of bodies is, but I rather describe a certain kind of being similar in every way to bodies, and whose creation we cannot deny to be within the power of God, so that we can hardly say that it is not body.[46]

Although Descartes's influence can be detected in this passage, there is also clear hostility to Cartesian positions. In so far as Newton sees the creation of matter not as the product of some necessity in nature, but as the result of divine will, and in so far as Newton makes use of the criterion of clear and distinct perception [*claram ac distinctam perceptionem*] there is a debt to Descartes. On the other side, in the denial of knowledge of the essence of body, and in the style of that denial, Newton places himself in strong opposition both to the method of discovery and to the first truth of the physical world presented in Cartesian metaphysics. That claim was that "the true nature of body consists solely in extension".[47]

Newton's mature account of the properties of bodies is found in Rule III of the *Regulae Philosophandi*, but the Rule contains no remarks about substance. In a draft version of the *Scholium Generale*, added to the second edition of the *Principia*, however, Newton expressed entirely Lockian sentiments:

> From phenomena we know the properties of things, and from the properties we infer that the things themselves exist and we call them substances: but we do not have any idea of substances. We see but the shapes and colours of bodies, we hear but sounds, we touch but external surfaces, we smell odours and taste flavours; but we know the substances or essences themselves by no sense, by no reflex action, and therefore we have no more idea of them than a blind man has of colours. And when it is said that we have an idea of God or an idea of body, nothing other is to be understood than that we have an idea of the properties by which bodies are distinguished from God or from each other. Whence it is that we nowhere argue about the ideas of substances apart from properties, and deduce no conclusions from the same.[48]

Whatever substance is, whether mental or physical, whether finite or infinite, although we know it exists, its essence is beyond our knowledge, Newton claims. All we may talk of with sense are properties for which we have corresponding ideas. A more Lockian position it would be difficult to imagine. But, whereas Newton's views, mostly unpublished, are largely offered unsupported, Locke reaches his conclusions only after sustained and detailed argument through over twenty pages.[49]

HYPOTHESES AND THE SCIENTIFIC ENTERPRISE

Locke was sceptical of the possibilities of scientific knowledge, "The systems of *natural philosophy* ... are to be read, more to know the *hypotheses* ... than with hopes to gain thereby a comprehensive, scientifical, and satisfactory knowledge of the works of nature", he wrote.[50] But he did allow that Newton's *Principia* was a success on a limited front. It had shown "how far mathematicks, applied to some parts of nature, may, upon principles that matter of fact justifie, carry us in the knowledge of some ... particular provinces of the incomprehensible universe".[51]

Hypotheses are uncertain, and hypotheses as speculative explanations should both be recognised as such, and also recognised as essentially an unsatisfactory way of proceeding in matters relating to natural philosophy. Locke's suspicion of hypotheses emerges frequently. The theory of innate ideas is an hypothesis, the view that a man always thinks is "a confused notion, taken up to serve an hypothesis".[52] Hypotheses too often lead to false explanations, and we are ever tempted to assume our hypotheses to be the truth. The hypothetical method has not proved successful in obtaining true knowledge. "He that shall consider how little general maxims, precarious principles, and hypotheses laid down at pleasure have promoted true knowledge", he wrote, including within this description, no doubt, Descartes and his followers, "will think we have reason to thank those who in this latter age have taken another course and have trod out to us, though not an easier way to learned ignorance, yet a surer way to profitable knowledge".[53] Here Locke would certainly have Boyle and Sydenham in mind, and probably Bacon as well. However, Locke was careful not totally to rule out hypotheses from science. The danger was always that we too readily accept an hypothesis because we are eager for explanation ...

We should not take up any one [hypothesis] *too hastily* ... till we have very well examined particulars and made several experiments in that thing we would explain by our hypothesis, and see whether it will agree to them all.[54]

An almost identical attitude is to be found in Newton's writings. Thus in his draft queries we find the following:

Could all the phaenomena of nature be (evidently) deduced from only three or four general suppositions there might be great reason to allow those suppositions to be true: but if for explaining every new Phaenomenon you make a new Hypothesis if you suppose y^t ye particles of Air are of such a figure size and frame, those of water of such another, those of Vitriol of such another, those of Quicksilver of such another ... If you suppose that light consists in such a motion precession or force & that its various colours are made by such & such variations of the motion & so of other things: your Philosophy will be nothing else than a systeme of Hypotheses. And what certeinty can be there in a Philosophy wch consists in as many Hypotheses as there are Phaenomena to be explained. To explain all nature is too difficult a task for any one man or even for any one age. Tis much better to do a little with certainty & leave the rest for others that come after, than to explain all Things by conjecture without making sure of any thing: And there is no other way of doing any thing, with certeinty than by drawing conclusions from experiments & phaenomena untill you come out general Principles & then from those Principles giving an account of Nature. Whatever is certain in Philosophy is owing to this method & nothing can be done without it.[55]

There are several points of note which arise in this passage. Newton, like Locke, does not totally dismiss the possibility that there may be some role for hypotheses in explaining nature. But the emphasis is clearly in another direction. Similarly, he is not at all attracted to the Cartesian ideal[56] of a completely deductive physics. Not that Descartes believed we could achieve that ideal. But he did hold it worthwhile to attempt to move as close as we could. Neither Locke nor Newton shared this aspiration. Indeed both saw in it the inherent danger of the temptation to take as certainties matters which were in truth only conjectures.

This hostility to the deductive ideal on the part of the two men did not lead to a position of scepticism. Between the twin dangers of aspiration to deductive science on the one hand and the nihilism of the sceptic on the other, there was a middle course which led to certainty of a kind which was appropriate to the condition of man. The certainty we can expect to obtain is that of inductive certainty, for the principles of natural science "are deduced from phenomena and made general by induction, which is the highest evidence that a proposition can have in this philosophy".[57] Universal proposition about the natural world were for Newton to be based on inductive evidence. Of course he recognized that their certainty was not guaranteed, but until there was evidence against such a generalization it should be accepted "as accurately or very nearly true, notwithstanding any contrary hypotheses that may be imagined, till such time as other phenomena occur by which they may either be made more accurate or liable to exceptions".[58]

Locke never set out any formal principles of method to correspond to those of the *Regulae*, but his position is undoubtedly that of Newton's. Mr. Newton's book can be accepted because it is based on principles "which matters of fact justify".

It is sometimes said that Locke believed that the ultimate truths of nature were necessary. There is certainly confused talk on his part about the status of laws of nature. But even if he does sometimes use language which suggests that he subscribes to a rationalist view,[59] such a position is totally inconsistent with a belief which both he and Newton certainly did hold, namely that the laws of nature were contingent upon the will of God. Locke makes a special point of contrasting the immutability of mathematical truths with truths about nature. The truths of geometry do not depend on any arbitrary power,

> But the coherence and continuity of the parts of matter; the production of sensation in us of colours and sounds, etc. by impulse and motion; nay the original rules and communication of motion being such, wherein we can discover no natural connexion with any *ideas* we have, we cannot but ascribe them to the arbitrary will and good pleasure of the Wise Architect.[60]

Locke supports this voluntarist position[61] by reference to miracles, which he points out, by argument *ad hominem*, none of his readers wishes to deny.[62]

Newton also placed the events of the world not in *necessity*, but in the *will* of God. The only route to knowledge of the world must be through experience because there is no other way whereby we may come to know the will of God in relation to the world. Thus Newton believed it was necessary to introduce direct intervention by God to account for the fact that from one original matter there should arise the luminous sun and the non luminous earth. This was an event not "explicable by mere natural causes, but I am forced to ascribe it to the counsel and contrivance of a voluntary Agent",[63] he wrote, and in the General Scholium he stated that "All that diversity of natural things which we find suited to different times and places could arise from nothing but the ideas and *will* (my emphasis) of a Being necessarily existing".[64]

The emphasis on the contingency of that which was discovered, but nevertheless the possibility of true discovery, is fundamental to the empiricist programme of Locke and Newton. In a sense it is its rationale. There is here no hint of Humean scepticism about the rationality of the programme. The contingency of knowledge was not a source of despair. It merely reflected a (contingent) fact about the relationship between man and God. For Locke it was part of the natural order of things. Man's place on the Great Chain of Being, and the knowledge which was thus available to him, was entirely appropriate to his station. For Newton, who, on occasion, saw himself as a child playing on the sea-shore when the whole ocean of truth lay undiscovered before him, was not thereby intimidated. Even if the inductive generalizations of the natural philosopher should turn out to meet with contrary instances, this would not be a hinderance to either knowledge or progress.

THE SYSTEM OF LOCKE AND NEWTON

It is now time to review the evidence assembled for the claim of the title of this paper. Let us begin with a terminological point. It might be said that the claim that Locke and Newton between them produced a system is necessarily misconceived because they were empiricists, and it is central to empiricism that there are no systems of philosophy and of the world; the empiricist just follows where the evidence takes him. In one sense clearly the point must be allowed. It must be allowed in precisely that sense in which we have argued that both Locke and Newton subscribed to the view that knowledge of the world may be gained only from experience, that the laws of nature are contingent, and that our claims to knowledge of the world may have to be revised in the light of experience. But there is another sense of system in terms of which the empiricist may be said to offer a system to the world. It is the sense in which we may and do correctly talk of the system of Aristotle. It is a system in the sense that, subscribing to an empirical source for the origins of our ideas, and an empirical method to attain knowledge of nature, the empiricist may yet produce a comprehensive, indeed a systematic, account of the world. Certainly there are gaps in the picture which Locke and Newton together offer (more on these below) but that they did produce a system in this sense can, I submit, be maintained. It might be outlined in the following way:

Man is a finite creature created by God with both physical and mental attributes. He has a mind and a body and he obtains knowledge of the world through the senses of the body which supply his mind with the basic materials for him to build up knowledge. He obtains knowledge of his own mind by internal reflection. Since his mind is limited to knowing only what his senses supply there are severe restrictions on what he can ever know. He can know that he has a mind and a body, but he cannot know the essence of these things. Indeed he cannot know the essence of any of God's creations, the substances of the world. He can, however, know the properties of those substances which come within the reach of his senses. The natural sciences are indeed the activity of collecting, ordering and inferring from those properties. Since the world is contingent on the will of God, and since our senses give us limited information, we can never be more than very confident that we have understood the world alright. Certainly we cannot expect to know anything about the world by *a priori* deduction. The correctness of this view of our knowledge of the world and its limitations on man is supported by its success in practice. The most able application of the method is to be found in the new science of Boyle, Sydenham, and especially Isaac Newton. The folly of the chief rival account is exemplified by the extravagant or demonstratively false claims of the Cartesians.

The work of Locke and Newton was to colour the approach to intellectual questions for at least a hundred years. The framework of knowledge created by Locke, and the exemplification of that framework within Newton's scientific writings have been powerfully and fruitfully influential ever since. Although between them their writings cover many major intellectual fields including metaphysics, logic, economics, theology, history, political theory, medicine, mechanics, mathematics and optics, their wide enquiries were set within the empirical context to which both subscribed and their conclusions limited by their empirical conception of proof. In matters of ethics, each saw the Bible as the primary source for our knowledge of right and wrong, and they saw the moral categories as a product of the will of God whose moral law, as Locke expressed it, we "may attain to the knowledge of, by the use and due application of our natural faculties".[65] Entirely at one with this Newton could write at the end of the *Opticks*: "And if natural Philosophy in all its Parts, by pursuing this Method [i.e. the experimental method as described by Newton] shall at length be perfected, the Bounds of Moral Philosophy will be also enlarged."[66] The empirical method not only has implications for science but for morals and religion also, and indeed, unsurprisingly, for all those areas of enquiry where both knowledge of man and knowledge of the world were required.

University of Keele

NOTES

* This is a longer version of a paper which was given to the symposium *Newton and the Enlightenment* at the Universita di Cagliari, Italy, in October, 1977.

[1] Locke has of course received much excellent treatment from historians of philosophy and intellectual historians. Philosophers, concerned more with refutation than accuracy, have not always been so kind. Locke's reputation among philosophers has tended to reflect the latest philosophical fashion, and historians of science have sometimes been content to follow the philosophers.

[2] Locke's manuscripts as a source for his involvement in science have been ably exploited in Maurice Cranston's *John Locke: A Biography*, (London, 1957), Kenneth Dewhurst's *John Locke (1632–1704) Physician and Philosopher. A Medical Biography* (London, 1963) and in numerous papers by individual authors. Newton's philosophical ideas have received important consideration from such scholars as I. B. Cohen, Alexandre Koyré, and J. E. McGuire.

[3] The material in this section draws heavily on two papers of mine 'Locke's *Essay* and Newton's *Principia*', *Journal of the History of Ideas* 37 (1978), 121–136, 'Locke, Newton and the Cambridge Platonists on Innate Ideas', *Journal of the History of Ideas*, forthcoming. I shall hereinafter refer to these papers as Rogers (1) and Rogers (2). Detailed substantiation of many of the points in this section may be found in them.

[4] Several examples of this position as held by noted scholars are given in Rogers (1) 121–122.

[5] For Locke's medical studies see especially Dewhurst, *op. cit.*, Note 2. For his involvement with Boyle see especially Dewhurst, 'Locke's Contributions to Boyle's Researches on the Air and on Human Blood', *Notes and Records of the Royal Society* 17 (1962), 198–206.

[6] There are several Locke notebooks extant which reveal his deep involvement in chemistry. Not only was he a diligent reader of the major texts, but also of the more obscure works. The notebooks also reveal that he conducted many experiments.

[7] On Stahl's chemistry classes see G. H. Turnbull: 'Peter Stahl, the First Public Teacher

of Chemistry at Oxford', *Annals of Science* 9 (1953), 265–270. The letters are to be found in *The Correspondence of Isaac Newton* (Cambridge, 1959–), III, letter numbers 382, 384, 389, 390, 391.

8 On this see the discussion in Rogers (1) 128–129.

9 See, for example, Gilbert Clarke's letter to Newton in September 1687, Newton *Correspondence* II, 485.

10 On this see James L. Axtell 'Locke's Review of the *Principia*', *Notes and Records of the Royal Society of London* 20 (1965), 152–161.

11 This Notebook, C.U.L. MS. 3996, has been discussed in several papers. See especially A. R. Hall: 'Sir Isaac Newton's Notebook', *Cambridge Historical Journal* 9 (1948), 239–250, and R. S. Westfall: 'The Foundations of Newton's Philosophy of Nature', *The British Journal for the History of Science* 1 (1962), 171–182.

12 J. E. McGuire is currently producing such an analysis.

13 There are several entires (not always attributed) from the writings of Henry More in the notebook.

14 For Barrow's methodological remarks, and particularly his strictures on hypotheses in natural science, see especially his *The Usefulness of Mathematical Learning Explained and Demonstrated: Being Mathematical Lectures Read in the Public Schools at the University of Cambridge*, translated by John Kirkby, (London 1734), Lecture VII.

15 *Physiologia Epicuro-Gassendo-Charltoniana* (1654) p. 3. I owe the observation that the heading of the *Questiones* is to be found in Charleton to Karin Figala.

16 On this see Rogers (1), 134.

17 For a discussion of this see Rogers (2). The remark is contained in a letter from Newton to Locke, Newton: *Correspondence* III 280.

18 Bodleian Library MS J. Locke c.5 f.62.

19 John Evelyn MSS, Christ Church College, Oxford, Letter Book V. I am grateful to Michael Hunter for this reference.

20 English translation of sixth edition, (1747), p. XXXVII.

21 *Ibid.*, p. XXXVIII.

22 It is of course true that not all Newton's expositors bothered with such explications. In neither the first, J. Keill's *Introductio ad Veram Physicam* (1701), nor, what is widely reckoned the best, Colin Maclaurin's *An Account of Sir Isaac Newton's Philosophical Discoveries* (1748), are they to be found.

23 This section draws particularly on my 'The Empiricism of Locke and Newton', Royal Institute of Philosophy Lecture, October, 1977, to be published in *Philosophers of the Enlightenemen* edited by S. C. Brown, (Harvester Press) forthcoming. I am grateful to the editor and the publisher for permission to incorporate material from that paper, which shall be referred to as Rogers (3).

24 See A. Koyré: 'Newton's *Regulae Philosophandi*' in his *Newtonian Studies* (London, 1965), especially 271–272, and I. B. Cohen: *Introduction to Newton's 'Principia'* (London, 1971) 30–31.

25 This translation is a modification of Koyré's translation, published in Koyré, *op. cit.*, 272. The original source is University Library, Cambridge, Add. MS. 3965.

26 See Rogers (2), where the comparisons between passages of the *Essay* and Newton's text are given in full.

27 There has been considerable confusion in the literature about both what Locke's views on space and time are, and also how they stand in relation to those of Newton.

Although in a very central sense Locke argues a relativist view of space and time those who can only see Locke as a disciple of Newton have actually attributed to him a commitment to Newtonian absolutes.

28 See, for example, the discussions in the following: J. A. Lohne: '*Experimentum Crucis*', *Notes and Records of the Royal Society* 23 (1968). R. S. Westfall: 'Newton and the Fudge Factor', *Science* 179 (1973); D. T. Whiteside, 'Newton's Lunar Theory: From High Hope to Disenchantment', *Vistas in Astronomy* 19 (London, 1976).

29 On the other side, he was often well aware that his own conjectures were *not* supported with sufficient evidence to justify their publication to the world, or even to be held as firm opinion by himself.

30 Both Locke and Newton probably owed as much to Descartes as they did to any thinker. Both also reacted critically to his stimulus and it is Cartesian views that both of their major works most strongly attack.

31 For discussions of Newton's speculations see J. E. McGuire: 'Force, Active Principles and Newton's Invisible Realm', *Ambix* 15 (1968); R. S. Westfall: *Force in Newton's Physics. The Science of Dynamics in the Seventeenth Century* (London, 1971) ch. VII.

32 On Newton's involvement in this project see Ralph W. V. Elliot 'Isaac Newton's 'Of an Universal Language', *Modern Language Review* 52 (1957), 1–28. See also Benjamin De Mott: 'Science versus Mnemonics. Notes on John Ray and on John Wilkins's *Essay towards a Real Character, and a Philosophical Language*', *Isis* 47 (1957), 3–13, and the same author's 'The Sources and Development of John Wilkins' Philosophical Language', *Journal of English and Germanic Philosophy* 57 (1958). See also Jonathan Cohen: 'On the Project of a Universal Character', *Mind* (1954), 49–63. Locke's argument in the *Essay* against the possibility of our having knowledge of the real essences of substances, and therefore of the possibility of there being a definitive system of classification of natural kinds and the natural order, was a powerful objection to the rationale of the programme associated with the *Characteristics Universalis*. Philip Sloan has shown that Locke's rejection of the possibility of obtaining knowledge of real essences of natural kinds had repercussions in science. See his 'John Locke, John Ray, and the Problem of the Natural System', *Journal of the History of Biology* 5, 1–53.

33 See *An Essay Concerning Human Understanding*, II, I, 1–5. All quotations will be taken from the *Clarendon Edition of the Works of John Locke*, edited by Peter H. Nidditch (Oxford, 1975). This edition is based on the fourth edition of the *Essay*, and will be cited by Book Chapter and Section numbers. Unless otherwise cited all Locke references are to the *Essay* in this edition.

34 I, IV, 18.

35 II, I, 9.

36 II, I, 18.

37 II, I, 21.

38 For some of the details see Rogers (3).

39 Newton to William Briggs, September 12 1682. *Correspondence* II p. 383.

40 As well as the letter already quoted see also the paper published in David Brewster's *Memoirs of Sir Isaac Newton* (Second edition, Edinburgh 1860) Vol. I, pp. 395–398.

41 U.L.C. Add. MS. 3996 f. 130.

42 *Ibid.*, f. 101.

43 *Ibid.*, f. 131.

44 *Ibid.*, f. 130.

[45] The date of this manuscript is disputed. The Halls place it about 1664. Others are inclined to place it considerably later. D. T. Whiteside, in a private communication, dates it as about 1671.

[46] *Unpublished Scientific Papers of Isaac Newton*, edited by A. R. and M. B. Hall (Cambridge, 1958) p. 138.

[47] *Principles of Philosophy* Part II, Principle V.

[48] U.L.C. Add. MS. 3965 f. 361–2. Published in Hall and Hall (*op. cit.* Note 40) p. 361.

[49] See especially *Essay* II, XXIII 'Of our Complex Ideas of Substances', *Passim.*

[50] *Some Thoughts Concerning Education*, in *The Educational Writing of John Locke*, edited by James L. Axtell (Cambridge, 1968), p. 305.

[51] *Ibid.*, p. 306.

[52] II, I, 18.

[53] IV, XII, 12.

[54] IV, XII, 13.

[55] U.L.C. Add. MS. 3970, f. 479.

[56] This ideal was expressed in a letter to Mersenne in 1640 thus: "I would think I knew nothing in Physics if I could only say how things could be, without proving that they could not be otherwise. This is perfectly possible once one has reduced everything to laws of mathematics; I think I can do it for the small area to which my knowledge extends. But I did not do it in my essays because I did not want to present my Principles there, and I do not yet see anything to persuade me to present them in future." Descartes, *Philosophical Letters* translated and edited by Anthony Kenny (Oxford, 1970) pp. 70–71. The *Principles* were in fact published four years later.

[57] In a letter to Roger Cotes. *Correspondence* V, p. 397. See also the draft letter to Cotes (pp. 398–9): "Experimental Philosophy reduces Phaenomena to general Rules & looks upon the Rules to be general when they hold generally in Phaenomena. It is not enough to object that a contrary phaenomenon may happen but to make a legitimate objection, a contrary phenomenon must be actually produced. Hypothetical Philosophy consists in imaginary explications of things & imaginary arguments for or against such explications, or against the arguments of Experimental Philosophers founded upon Induction. The first sort of Philosophy is followed by me, the latter too mucy by Cartes, Leibniz & some others."

[58] Rule IV *Regular Philosophandi, Principia* Book III. Andrew Motte's translation revised by Florian Cajori, (Berkeley and Los Angeles, 1962) p. 400.

[59] R. S. Woolhouse in his *Locke's Philosophy of Science and Knowledge*, (Oxford, 1971) attribute rationalism to Locke, which he defines as the view "that in natural science we have to do with synthetically necessary propositions which are not by us knowable *a priori*" (p. 25).

[60] IV, III, 29.

[61] On the voluntarist tradition in this context see Francis Oakley: 'Medieval Theories of Natural Law: William of Ockham and the Significance of the Voluntarist Tradition', *Natural Law Forum* 6 (1961).

[62] Locke expresses a similar position in his *A Discourse of Miracles*, published posthumously in 1706.

[63] *Correspondence* III, p. 234. Newton to Richard Bentley, 1692.

[64] *Principia* (*ed. cit.* Note 65) p. 546.

[65] *Essay* I, III, 13.

[66] *Opticks* (4th edition, London 1730). Dover edition, p. 405.

INDEX

STUDIES IN THE HISTORY OF MODERN SCIENCE

Editors:

ROBERT S. COHEN (Boston University)
ERWIN N. HIEBERT (Harvard University)
EVERETT I. MENDELSOHN (Harvard University)

1. F. Gregory, *Scientific Materialism in Nineteenth Century Germany*. 1977, xxiii+279 pp.
2. E. Manier, *The Young Darwin and His Cultural Circle*. 1978, xii+242 pp.
3. G. Canguilhem, *On the Normal and the Pathological*. 1978, xxiv+208 pp.
4. H. C. Kennedy, *Peano*. 1980, xii+230 pp.
5. L. Badash, J. O. Hirschfelder, and H. P. Broida (eds.), *Reminiscences of Los Alamos, 1943–1945*. 1980, xxi+188 pp.
6. L. L. Bucciarelli and N. Dworsky, *Sophie Germain: An Essay in the History of the Theory of Elasticity*. 1980, xi+147 pp.
7. D. J. Struik, *The Land of Stevin and Huygens*. 1981, xiv+162 pp.
8. I. Langham, *The Building of British Social Anthropology*. 1981, xxxii+392 pp.
9. Z. Bechler (ed.), *Contemporary Newtonian Research*. 1982, vi+241 pp.
10. W. T. Sullivan, III, *Classics in Radio Astronomy*. 1982, xxiv+348 pp.
11. L. Anderson, *Charles Bonnet and the Order of the Known*. 1982, xvi+160 pp.